The Ethics of Screening in Health Care and Medicine

INTERNATIONAL LIBRARY OF ETHICS, LAW, AND THE NEW MEDICINE

Founding Editors

DAVID C. THOMASMA†
DAVID N. WEISSTUB, *Université de Montréal, Canada*
THOMASINE KIMBROUGH KUSHNER, *University of California, Berkeley, U.S.A.*

Editor

DAVID N. WEISSTUB, *Université de Montréal, Canada*

Editorial Board

TERRY CARNEY, *University of Sydney, Australia*
MARCUS DÜWELL, *Utrecht University, Utrecht, the Netherlands*
SØREN HOLM, *University of Manchester, United Kingdom*
GERRIT K. KIMSMA, *Vrije Universiteit, Amsterdam, the Netherlands*
DAVID NOVAK, *University of Toronto, Canada*
EDMUND D. PELLEGRINO, *Georgetown University, Washington D.C., U.S.A.*
DOM RENZO PEGORARO, *Fondazione Lanza and University of Padua, Italy*
DANIEL P. SULMASY, *The University of Chicago, U.S.A.*

VOLUME 51

For other titles published in this series, go to
http://www.springer.com/series/6224

Niklas Juth · Christian Munthe

The Ethics of Screening in Health Care and Medicine

Serving Society or Serving the Patient?

Niklas Juth
Karolinska Institutet
Dept. Learning, Informatics, Management
  & Ethics (LIME)
Berzelius väg 3
171 77 Stockholm
Sweden
niklas.juth@ki.se

Christian Munthe
University of Göteborg
Dept. Philosophy, Linguistics &
  Theory of Science
PO Box 200
405 30 Göteborg
Sweden
christian.munthe@gu.se

ISSN 1567-8008
ISBN 978-94-007-2044-2   e-ISBN 978-94-007-2045-9
DOI 10.1007/978-94-007-2045-9
Springer Dordrecht Heidelberg London New York

Library of Congress Control Number: 2011936796

© Springer Science+Business Media B.V. 2012
No part of this work may be reproduced, stored in a retrieval system, or transmitted in any form or by any means, electronic, mechanical, photocopying, microfilming, recording or otherwise, without written permission from the Publisher, with the exception of any material supplied specifically for the purpose of being entered and executed on a computer system, for exclusive use by the purchaser of the work.

Printed on acid-free paper

Springer is part of Springer Science+Business Media (www.springer.com)

Acknowledgements

This book is the result of work carried out over a period of at least ten years, probably longer. The awareness of the importance and variety of ethically relevant ways in which screening programmes differ from other ways of organising the offering and use of medical testing methods has been growing gradually in the context of both of our past research on the ethics of, e.g., genetic testing, prenatal diagnosis, reproductive medicine and public health. The actual idea of writing a book specifically about the ethics of screening was, however, a slightly more recent one, and in a way it came about due to a mistake. In late 2004, we were asked by the editors to contribute a chapter on the ethics of screening to the 2nd edition of the massive collection *Principles of Health Care Ethics*[1] (a classic advanced reader and reference work in the fields of bioethics and general health care as well as medical ethics). Flattered, we accepted and got to work without checking properly what was actually expected of us, with the result that we submitted a manuscript of about 22,000 words and were harshly told to cut it down to about 5000 (which we did, of course[2]) with the off-hand remark that we could always use the longer text as the basis for a book. Looking back, we are extremely grateful to Richard Ashcroft, Angus Dawson, Heather Draper and John McMillan – both for steering us onto this path in the first place and for stimulating our motivation for our subsequent work by applying strict editorial discipline to our tendency of enthusiastically getting ahead of ourselves. We are equally grateful to David Weisstub, who, over a nice cup of coffee in Stockholm autumn 2009, took an interest in the manuscript and suggested we send it to Springer and the International Library of Ethics, Law, and the New Medicine Series.

---

[1] Ashcroft et al. (2007).
[2] Juth and Munthe (2007).

Research – even in ethics – and writing furthermore requires time and resources and we have been lucky to have been awarded that in the context of a series of projects where themes appearing in this book have been overlapping in various ways. The Swedish Ethics in Health Care Programme generously funded the project *Presymptomatic Testing and Genetic Counselling: Goals and Ethics for Clinical Practice, Caring and Education* (2000–2008), as did the European Commission of *European Public Health Ethics Network (EuroPHEN)* (2002–2006) and, later, the Swedish Government and the University of Gothenburg funded projects and collaborations from 2006 that eventually has become the *University of Gothenburg Centre for Person Centred Care*. Karolinska institute generously funded more time for research for one of the authors, funds without this book would not have been completed.

Already when working on the original chapter, we received valuable assistance of several people, besides the above-mentioned editorial team, and this aspect of the work has, of course, expanded as we set out to realise the vision of a book. Ethics researchers are at heart theoretical analysts, so when ethics is applied to concrete areas of practice one becomes extremely dependent on experienced and wise guides and door openers with regard to how this practice actually works and reliable information and pedagogical explanations of the many scientific aspects that are relevant. Accordingly we would like to thank The-Hung Bui (senior clinical consultant of clinical genetics, prenatal diagnosis and foetal medicine at Karolinska Institute), Ulrika von Döbeln (head of the Swedish neonatal screening programme, Karolinska Institute), Karl-Henrik Gustavson and Jan Wahlström (retired professors of clinical genetics at Uppsala University and the University of Gothenburg, respectively), Darren Shickle (professor of public health at the University of Leeds), Erik Björck (chief of medicine in clinical genetics, Karolinska institute) and Amy J Hoffman (project manager of the Newborn Screening Translational Research Network at the American College of Medical Genetics & ACMG Foundation), as well as the working group on bioethics and legal issues of the same institution. Partly in the context of the projects mentioned above, we have received further assistance in the context of discussions by numerous colleagues in our own field, where we (besides the mentioned editorial team) accidentally and regretfully happen to remember only Henrik Ahlenius, Bengt Brülde, Daniela Cutas, Gert Helgesson, Göran Hermerén, Søren Holm, Matti Häryry, Niels Lynöe, Lars Sandman, Manne Sjöstrand, Tuija Takala, and Maja Wessel. Two reviewers

Acknowledgements

for Springer – one of which, Marcel Verweij, chose to waive anonymity – took their job admirably serious and provided a great number of perspectives, critical points and pieces of information without which this book would have been much inferior to the present version.

Needless to say, none of the people or organisations mentioned are to be held accountable for any of our remaining mistakes and flaws.

Contents

1 Introduction 1
   1.1 The Wilson and Jungner Criteria 3
   1.2 Plan and Point of the Book 5
   1.3 The Concept of Screening 6

2 Why Screening? 13
   2.1 Screening, Treatment and Prevention: Preliminary Remarks . 14
   2.2 Health: Life and Well-Being 18
      2.2.1 Health and Counselling 19
      2.2.2 The Good of People and of the Population 20
   2.3 Autonomy 22
      2.3.1 Respecting and Promoting Autonomy 23
      2.3.2 Promoting and Respecting Autonomy
            Through Screening 25
   2.4 Justice 27
   2.5 Summary 29

3 Screening – What, When and Whom? 31
   3.1 Diseases and Groups 31
      3.1.1 Prenatal Screening 33
      3.1.2 Neonatal Screening 42
      3.1.3 Child and Adolescent Screening 53
      3.1.4 Adult Screening 58
   3.2 Investigation, Testing and Analysis 60
      3.2.1 Safety 61
      3.2.2 Validity 63
      3.2.3 Predictive Value 67
   3.3 Treatments 72
      3.3.1 Abortion as a Treatment 73
      3.3.2 Counselling as a Treatment 75
   3.4 Summary 78

4 Screening – How? ... 81
- 4.1 Informed Consent ... 82
- 4.2 Counselling ... 87
  - 4.2.1 Genetic Counselling as a Template ... 88
  - 4.2.2 Expansion: Shared Decision Making ... 93
- 4.3 Funding and Participation ... 95
- 4.4 Summary ... 97

5 Case Studies ... 99
- 5.1 Non-invasive Prenatal Diagnosis ... 100
- 5.2 Neonatal Screening for Fragile X ... 108
- 5.3 Mammography Screening ... 114
- 5.4 PSA Screening for Prostate Cancer ... 122

6 Serving Society or Serving the Patient? ... 127
- 6.1 Summary of the Analysis so Far ... 127
- 6.2 The Public Health – Health Care Tension Area ... 130
- 6.3 The Relevance of a Social Science Perspective ... 132
- 6.4 An Institutional Approach to Health-Related Ethics: A Sketch ... 135
- 6.5 Applying the Institutional Approach: Three Cases ... 137
  - 6.5.1 Institutions, Functions and Ethics: Prenatal Care vs. Communicable Disease ... 138
  - 6.5.2 Direct to Consumer Genetic Testing: The Limits of Context Relativity ... 141
  - 6.5.3 Screening and Justice: When to Spend Health Care Resources on Screening ... 147
- 6.6 Revisiting the Wilson and Jungner Criteria for Screening ... 149
- 6.7 Closing ... 157

References ... 165

Index ... 175

Chapter 1
Introduction

Medical screening programmes are amongst the most debated aspects of health care practices in medical ethics as well as health policy discussions. There are many explanations for this, but a chief one is the fact that screening programmes affect large numbers of people. This, in turn, connects to the history of screening, which is strongly linked to the development of the area of public health (as opposed to individual health care) during the twentieth century, and thus to the use of medical knowledge and technology for societal aims that transcend those considerations that arise in the context of a health care professional interacting with an individual patient.[1] For this reason, screening is an interesting case for the study of what becomes of medical and health care ethics in situations where medical professionals act primarily as servants of society. However, it also serves to partially explain the nature of the ethical controversies surrounding screening and to highlight gaps in state of the art health care ethics perspectives that may be filled by ethical deliberation based on a public health perspective.

Qualms about the various ways in which health care and medicine through modern history have acted primarily or partially as servants of society (areas like population control, criminal law, eugenics, military endeavours, et cetera, come to mind) have been a driving force in post World War II medical ethical debates. Equally so, the results of these debates in the form of ethical principles and regulation aimed at protecting individuals from being sacrificed in the name of some overarching societal objective within the medical research and health care contexts. At the same time, most medical professionals, as people in general, recognise the importance of public health also from a health care perspective, since the idea of a functioning health care system without a backdrop of successful public health measures seems plainly unrealistic. In recent years, this general recognition has been

---

[1] See, e.g., Dawson and Verweij (2007a).

attracting more attention from medical professionals as well as policy makers due to pandemics scenarios, political debates regarding the financing of health care systems, e.g. in the USA, challenges faced by developing countries when trying to establish welfare societies and increased general focus on global health issues, to name a few aspects. Still, the initial ethical tension between public health and health care perspectives on health policy issues remains and is a main theme in the relatively new field of public health ethics research.[2]

Another aspect that underlines the need for ethical reflection is that there are a number of social and economic aspects that make screening exercise a particular pull on health care professionals and institutions. As will be explained in greater detail, a screening programme is an ambitious undertaking requiring not only significant amounts of manpower and technical resources, but also that these are structured into a centralised organisation. This means not only that screening is a costly effort (which we will return to), but also that being in charge of a screening programme is an attractive prospect, be it for pure business reasons (regarding private health care and health related business), increased professional status, or access to resources that may be utilised for other objectives (such as basic research). On top of that, the general logic of organisations to tend to emphasise reasons for accepting whatever suggestion that will expand the organisation must be taken into account.

There is also reason to believe that screening will increasingly be on the agenda in medical ethics. Although screening programmes may target very different diseases and employ a host of different detection methods, the development of genetic knowledge and technology in recent decades has opened up a large area of hitherto unconsidered possibilities for screening that has already provoked debate. It seems plausible to expect that this development will continue.

In addition, the general resurgence of the attention to public health considerations within medicine in the wake of new risk scenarios for infectious disease (pandemics as well as antibiotic resistant strains of bacteria), health policy of developing countries, global poverty and health inequalities mentioned above can be complemented by further developments besides genetics. The general tendency in developed societies to "medicalise" more and more aspects of life, such as mood variations, stress, criminality,

---

[2] Childress et al. (2002); Dawson (2011); Dawson and Verweij (2007b); Munthe (2008); Rhodes (2005); and Shickle et al. (2007).

alcohol and drug consumption and sport, can and have in some instances been proposed or even applied as an adequate basis for medical testing programmes.[3] At the same time, longstanding and, in public health terms, markedly successful screening efforts – such as publicly run routine developmental checkups of children and adolescents or general health checkups for older adults – currently face challenges, both through the strains on national health services and through criticism based on the basically individualist tenets of tradition health care ethics.

For all these reasons, then, the ethical aspects of screening should be considered a central issue of all areas of health and health care related ethics for quite some time and may, in extension, serve as a primary case for advancing approaches to these areas of ethics that take into account the role of medicine and health care seen from a more overarching societal point of view.

1.1 The Wilson and Jungner Criteria

In the 1960s, medical developments resulted in increased possibilities to rapidly and cheaply test populations for risks of disease, which made screening an area of controversy. Because of this, the World Health Organisation (WHO) asked Gunnar Jungner and James MG Wilson to write a report with guidelines for the introduction of screening programmes. Their report, *Principles and practice of screening for disease*, was published in 1968 and is, as a matter of fact, still to this day considered as the "gold standard of screening assessment".[4] Although Jungner and Wilson themselves explicitly stated that they did not consider the ten criteria they formulated in the report to be the last word on the issue of which screening programmes are warranted, later attempts to formulate timely principles for assessing screening programmes have as a rule taken the Wilson and Jungner criteria as a point of departure and merely, more or less, reformulated them.[5] This demonstrates the influence and strength of the criteria not only as tools of health care policy making, but suggests that they, in fact, contain substantial ideas that have stood the test of time. Certainly, therefore, a comprehensive book on the ethics of screening cannot bypass them.

---

[3] See, e.g., Chalmers (2005); McNamee et al. (2009); Munthe (2005); and Savulescu (2005).
[4] Linnane et al. (1999).
[5] See e.g. WHOs latest contribution in this area (Andermann et al. 2008).

The ten "commandments" of screening set out by Wilson and Jungner are the following:

(1) The condition sought should be an important health problem.
(2) There should be an accepted treatment for patients with recognized disease.
(3) Facilities for diagnosis and treatment should be available.
(4) There should be a recognizable latent or early symptomatic stage.
(5) There should be a suitable test or examination.
(6) The test should be acceptable to the population.
(7) The natural history of the condition, including development from latent to declared disease, should be adequately understood.
(8) There should be an agreed policy on whom to treat as patients.
(9) The cost of case-finding (including diagnosis and treatment of patients diagnosed) should be economically balanced in relation to possible expenditure on medical care as a whole.
(10) Case-finding should be a continuing process and not a "once and for all" project.[6]

We will repeatedly refer to and discuss these criteria in the course of the book. However, as ethicists we want to point out one particular reason for why a book is indeed needed, rather than presenting yet another variation on the Wilson and Jungner theme: These principles themselves do not contain or suggest their own *rationale* or any idea as to on what grounds they should be preferred over alternative ones. Due to this, the principles themselves also provide no guidelines for how different interpretations of the principles should be assessed, and how resulting ethical, practical and scientific conflicts about suggested screening programmes should be resolved. In this book, we will try to provide such an overarching rationale; we will try to identify which factors are important in deliberating about existing or proposed screening programmes and address issues about how to reason when different factors seem to pull in opposite directions. In the final chapter we will present an evaluation and discussion of the Wilson and Jungner criteria in the light of our results (see Section 6.6). The conclusion of this discussion, as indicated, is that criteria indeed deserve their special standing and still provides a good point of departure for assessing screening programmes. However, they need to be complemented, interpreted, and made more precise. And to be able to say *how* they need to be so amended, we have

---

[6] Wilson and Jungner (1968).

found it necessary to venture into the various complicated ethical discussions undertaken throughout this book.

1.2 Plan and Point of the Book

In this book, we will give an introduction to and analyse the ethical aspects of screening. A main aim is, of course, to make the chief questions clear for further investigation and discussion. However, analysing these issues further and suggesting solutions to some of the problems found is also among our objectives. Obviously, the notion of ethics employed in this endeavour expresses a more inclusive concept than the one put to use when discussing, e.g., moral problems facing health care professionals. It is in the nature of screening to occupy an area that overlaps the health care sector and other societal enterprises, as well as connect to the general political interest in the length, functionality and quality of life of citizens. Ethics with regard to screening thus has to include normative and value issues arising from these other points of view.

In the next section of this introductory chapter, a clarification of the concept of screening designed to serve the purpose of ethical inquiry will be offered. The following chapter will then move on to the basic issue of what values may be promoted by screening programmes. Considering these values, we will also point to how they may sometimes conflict and how failures to promote them may be a drawback of a screening programme. In Chapter 3, we will then consider the more detailed issue of what aspects of a screening programme that are relevant for determining whether it should be undertaken or not. In that context, various ways of delimiting the targeted disease or population are considered, but also issues regarding the quality of testing methods and follow-up procedures. Given that a screening programme is considered defensible to undertake as such, the further question arises of how its implementation should be organised in order for the programme to remain defensible. This is the issue discussed in Chapter 4, where ideas about the importance of counselling and consent are central, but where the question of using incentives to boost the uptake of the programme will also be analysed. In Chapter 5, we apply the lessons from the previous chapters in four case studies considering currently contested and possibly forthcoming controversial screening programmes. The cases are: non-invasive prenatal diagnostic screening, neonatal screening for fragile x, mammography screening for breast cancer and PSA screening for prostate cancer.

The conclusions of the discussion regarding screening as such is summarised in Chapter 6, and thereafter used as a basis for expanding the discussion to consider the nature of medical and health care ethics with regard to activities and practices that are, at least partly, undertaken as a service to society. The ultimate points argued for are that the case of screening serves to illustrate that health care and medicine, as a matter of fact, normally involve quite a lot of such areas, that issues arising in these areas call for a higher degree of integration between ethics and social science perspectives on health care and medical research, and that the traditionally individualistic and profession-focused perspective of medical and health care ethics therefore needs to be supplemented by an institutional approach.

1.3 The Concept of Screening

Since screening is a well-established ingredient in health care and health policy, it is a bit surprising to find that there is no unanimously accepted standard definition of the term "screening" in medicine, health care or related research areas. Moreover, many of the definitions that have been proposed are not specifically tailored to the needs of an ethical analysis. For this reason, we will spend some effort on discussing the very concept of screening as a basis for presenting the definition that will be applied in this book.

The utilisation of medical investigation and testing methods is, of course, one necessary feature of the concept of screening, but such methods are used also in contexts not possible to characterise as screening – for instance, the diagnostic investigation of an individual patient in the ordinary health care setting. When trying to single out what is special about screening, nowadays, some of the following further features are often mentioned: screening aims at *selecting individuals* at risk of disease(s) from a (large) *population* of individuals *not united by previously recognized risk or symptoms* of the disease(s) in question by relatively *rapid and cheap* means. Moreover, there seems to be reasonable consensus on what constitutes typical examples of screening: routine ultrasound in pregnancy care, neonatal testing for phenylketonuria (PKU) and similar diseases, mammography breast cancer-testing programmes, et cetera.

In effect, screening needs to be characterised in terms of *a certain way of organising the offer and use of medical testing methods in the pursuit of a certain aim*. The question then is: what way and what aim?

In their classic text, Wilson and Jungner emphasised screening as a procedure that primarily aims at selecting individuals for further investigation.[7]

---

[7] Wilson and Jungner (1968, p. 11) thus underlines that a "…screening test is not intended to be diagnostic. Persons with positive or suspicious findings must be referred to their physicians for diagnosis and necessary treatment."

1.3 The Concept of Screening

Screening in this sense is a kind of filtering device, sorting out those who probably have or will have some disease(s) from a larger group. Due to the rapid development of genetics, genetic testing techniques and the resulting focus in the debate on genetic screening, the emphasis on the distinction between screening and clinical or diagnostic investigation has, however, decreased significantly.[8] In genetic screening, the methods used on individuals for diagnosing or determining risk of disease and those utilised in screening programmes are often the same. The general idea of screening as a "filter" applied at a population level to select sub-groups of this population for further action nevertheless remains.

Instead of the contrast between the techniques applied in screening as compared to other contexts, today, the difference regarding the direction of the initial contact typical to a screening programme is often emphasised: Typically, in such a programme, the *initiative to the investigation is not coming from the investigated individual herself*, but rather from health care or society.[9] One reason for bringing out this feature as typical is that many ethics and policy issues concerning actual screening programmes arise due to the fact that testing through screening is not initiated by the individual being tested. This means that the typical feature leading up to health care measures being undertaken – the event that an individual perceives a health problem and therefore seeks out health care – is not present in a screening programme. Quite the opposite, it is health care, often as a servant of some societal aim, which seeks out the individual. On the one hand, this has been presented as an advantage of screening: people who would otherwise have failed to take initiatives in order to be tested, perhaps due to ignorance, are given an opportunity to gain from the possible benefits of being tested, e.g. from treatment or reassurance. Therefore, screening programmes can and have been defended on grounds of public health and justice.[10] On the other hand, the initiative not being the individual's own has its downsides, for instance, the risks of reducing autonomy and provoking unnecessary anxiety. This is so, partly since the fact that the initiative comes from someone else than the individuals themselves means that these individuals seldom have any suspicion or worry of being at risk of disease to start with.

The idea of screening as a filtering and selection procedure presupposes that screening is directed towards a more or less *large population of*

---

[8] Although the characteristic of screening as a selection procedure is sometimes mentioned also in relation to genetic testing (Shickle, 1999, p. 1).
[9] The Danish Council of Ethics (1999), chapter 2.2.
[10] Hoedemaekers (1999), pp. 209–211.

people. This supposition is often made explicit in definitions of screening,[11] although professionals sometimes use the notion screening in other ways.[12] However, since the latter form of usage is almost absent in ethical discussions, we will stick to the more common conception that presupposes or explicitly states that screening concerns large populations of people.

It should nevertheless be observed that there is no precise limit to *how* large such a population has to be in order for the concept of screening to be applicable. In some cases, screening may concern all people in a country or a multi-national region, or large segments of such a group, such as all men, women or children. In other cases, the geographical scope may be more narrow, or the population targeted may be more specific with regard to some condition thought to be relevant from a health perspective, such as being pregnant, being within a certain age span, belonging to a particular ethnic group, having a history of certain health problems, et cetera. However, screening must, at the same time, be held separate from the sort of register-based routine monitoring of the health of the population that takes place, for example, as a standard ingredient of the public health measures of most developed countries. Such programmes do not involve the direct use or offer of using medical investigation or testing methods, but rather assemble data from activities of various sorts (screening programmes as well as general health care diagnostics) where such methods have been applied. Even taking this distinction into account, however, the feature of screening to be directed at some large population seems necessarily vague.

Further imprecision of the concept of screening results from the fact that the property of screening of targeting individuals *not united by previously recognized risk or symptoms* has to be understood as a matter of degrees. There are screening programmes where the population screened is not united by any recognized increase of risk at all, for instance the neonatal screening programmes running in many countries.[13] These screening programmes may thus be described as *pure*. However, many proposed and performed screening programmes are directed towards populations where there is or may be some previously recognized risk of disease. One type of example is genetic carrier screening programmes confined to certain ethnic groups known to have an increased risk compared to other populations, e.g. carrier

---

[11] See e.g. Kinzler et al. (2002), p. 277; and Shickle (1999), p. 1.

[12] It is, for example, commonplace in the medical literature to find talk about screening cells (when analysing them in a laboratory), of screening the organs of one single individual (e.g. in obstetric ultrasound), or even when referring to the application of any testing procedure on a single person in order to perform a diagnosis or risk assessment.

[13] Chadwick et al. (1999).

1.3 The Concept of Screening

screening for thalassemia among people of a Mediterranean origin,[14] for sickle cell anaemia among people of African origin,[15] for Tay-Sachs among Ashkenazi Jews,[16] or for cystic fibrosis among Caucasians.[17] Another example is prenatal screening programmes for chromosomal disorders, e.g. Down syndrome, directed towards pregnant women above a certain age.[18]

The previously recognized risk of disease may be even more obvious if one includes cases like those where one person is tested positively in an ordinary health care setting and others are approached as a result of this. One example of this concerns communicable diseases, such as HIV, where one positive finding can lead health care to contact others at risk. Another example is so-called cascade genetic testing, i.e. when health care initiates an investigation on someone's relatives based on a positive test result of that individual.[19] The common feature of these examples is once again that the individuals approached need not have any initial suspicion of being at risk of some disease, but the likelihood that they actually do in e.g. cascade genetic testing is bigger than in pure screening. Below, we will focus on screening programmes directed towards larger populations than normally is the case regarding cascade testing (which are most often confined to specific biological families), since the problems of screening then are more salient. With this being said, some of the arguments are relevant also regarding cases involving fewer individuals.

Screening is traditionally thought of as a tool for detecting and treating disease in its early stages or, preferably, preventing disease before onset.[20] This is why screening primarily regards individuals who have no previous suspicion of being at risk of disease. Therefore, some definitions of screening explicitly mention early detection as a purely conceptual feature.[21] However, strictly speaking, to emphasise early detection seems to be a way of highlighting a primary *reason* for screening rather than defining what screening is. In an ethics context, it is desirable to keep these two aims of

---

[14] Ioannou (1999).
[15] Marteau and Richards (1996).
[16] Sutton (2002).
[17] Gregg and Simpson (2002).
[18] Chadwick et al. (1999).
[19] This has been called an "inwards-out" approach to screening (Shickle and Harvey, 1993).
[20] So-called secondary and primary prevention, respectively (Wilson and Jungner, 1968, p. 14; and Shickle, 1999, p. 1).
[21] ESHG (2003), p. 5.

selecting a definition apart (in order not to beg substantial ethical issues). The *concept* of screening should thus give room for bad as well as good cases of screening. An *ethics* of screening will then use this concept for formulating ideas about what types of screening *are* good and bad.

Some definitions of screening also include the condition that the selection procedure should be rapid and/or (relatively) cheap.[22] This relates to several of the characteristics mentioned above, for instance the idea of screening as a primary sorting mechanism for further, more careful diagnosis and the notion of screening as being primarily targeted at large populations. Of course, it is also related to the commonly stated purpose of screening as a preventive public health measure. Some kind of cost-benefit analysis underlies any such measure[23] and, thus, "[e]conomic balance of the cost of case-finding in relation to total expenditure on health care"[24] is an important aspect of the justification of screening – requiring analysis of the costs and benefits of screening programmes in relation to other parts of health care and health policy. Since the benefits are often cast in terms of promoting health, well-being or autonomy, and the costs, besides economic, regard the reduction of such values, many arguments for and against screening can be analysed in a consequentialist framework.[25] Thus, although the test being cheap and rapid is often seemingly brought out as a defining feature of screening, it should primarily be understood as a reminder of the importance of balancing various costs and benefits. However, just as in the case of early detection, this has more to do with ideas about what may *justify* a screening programme than what makes it a *screening programme* in the first place. It should be noted, though, that this aspect of what may make screening good or bad serves to highlight further ethical problems typical to screening: if the testing procedure is a relatively rapid and cheap one, the problem of mistaken results tends to become more acute,[26] as does the potential problem of lack of proper counselling or information.[27]

Summing up our conceptual point of departure, then, we will in the rest of this book use the following working definition of the term "screening":

> The use of medical investigation or testing methods at the initiative of health care or society for the purpose of investigating the health status of individuals with the aim of selecting some of these for possible further treatment from a large

---

[22] Wilson and Jungner (1968, p. 11); and Shickle (1999, p. 1).
[23] Kinzler et al. (2002).
[24] Wilson and Jungner (1968), pp. 35–37.
[25] Hoedemaekers (1999), p. 224.
[26] See Section 4.2.
[27] Shickle (1999), p. 11.

1.3 The Concept of Screening

population of people that is not united by previously recognised risk or symptoms of disease.[28]

What we will be doing in the rest of this book, then, is to consider ethical issues specific to screening in this particular sense. We recognise, however, that several of these issues thus discussed may be applicable also in cases where the screening is not as pure, for example, where the population is not that large, when there is some initial awareness of risk of disease, or when the party taking the initiative is someone else than health care or society. Some of these "impure" variants will be considered continuously through our discussion, while others will be elaborated on in the final chapter.

---

[28] In this definition, we are presupposing a broad sense of treatment, including e.g. dietary recommendations or counselling (see Section 2.1).

Chapter 2
Why Screening?

In order to determine whether a certain screening programme should be implemented or not, one has to ponder the basic issue of why such an effort may be worthwhile at all. That is, in order to provide a rationale or justification for any screening programme one has to have some idea about the values that the programme should promote. Of course, whatever values a certain screening programme may promote, this promotion has to be balanced against the potential drawbacks of the programme. Thus, as already mentioned, much ethical debate about screening is cast in terms of some kind of cost benefit-analysis. It must be noted, though, that in the context of an ethical analysis, terms such as "cost" and "benefit" must be understood very broadly. They can only partly be translated into straightforward monetary figures, since the values and drawbacks often regard qualities mentioned in basic ethical theories, like autonomy, justice, or well-being. Moreover, unlike monetary costs, it cannot be taken for granted that these values can or should be traded off against each other in any way imaginable. To substantiate more particular ideas on what values may motivate what costs or sacrifices in terms of other values and in what way, explicit (and thereby controversial) moral premises need to be a part of the argument.

In this chapter, the most common general basic values used to justify screening will be presented.[1] We will also describe some of the conflicts with respect to these values. In addition, drawbacks or "disvalues" in the light of these values specific and common to all screening programmes will be described. This overview paves the way for a more detailed discussion (in Chapters 3, 4, and 5) regarding what screening programmes are justified and how they should be implemented. The general idea behind this

---

[1] We accept that these basic values really are valuable. If one does not presuppose that, most notably, well-being and/or autonomy are valuable in the first place it becomes difficult to see how one could justify any health care at all.

strategy of approaching the analysis is that if one can say something substantial about why screening programmes should or should not be considered in the first place, one can then use this as a basis for evaluating which programmes that actually promote the proposed values to a sufficient degree. However, we will in the present chapter *not* present any sort of final verdict at to what value or goal is more important when conflicts between different goals/values occur. Rather, the outcome of the discussion in this as well as the later chapters is that the idea of substantiating such a verdict is a very complex and demanding undertaking that requires some theoretical innovation. In the final chapter, we outline one suggestion for how that challenge may be met by further research.

2.1 Screening, Treatment and Prevention: Preliminary Remarks

As already mentioned, the most classic idea of what may make screening a good thing has been formulated in the context of public health, the classic goal of which is the promotion of *population health* – i.e. the aggregated health levels of the individual members of a population.[2] In order for population health in this traditional sense to be promoted by a screening programme, there needs be some kind of *acceptable treatment* attached to the programme. Wilson and Jungner held this criterion to be about a treatment for the condition screened for and considered it to be the most important one to be met in order for a screening programme to be justified.[3] Some preliminary analysis of what is implied by this condition will serve as a stepping-stone to the inquiry into what values that may be served by screening and how. Further, more specific, discussion of the treatment condition will be undertaken in Section 3.3.

It is clear that the notion of treatment here cannot be restricted to *reparative* efforts – measures applied to achieve cure or, at least, damage control or amelioration of discomfort, in the case where a person is already suffering from some damage or disease. On the contrary, the treatments actualised

---

[2] There are quite a few ideas about both what is to be included in the concept of health, how an aggregation of individual health levels should be structured in order to reflect the health of the population and what empirical factors are relevant for this (Murray et al., 2002). Still, in practice, standard measurements applied assume that things like average morbidity, mortality (especially infant mortality) and life expectancy are chief determinants of the health of a population.

[3] Wilson and Jungner (1968), p. 27.

2.1 Screening, Treatment and Prevention: Preliminary Remarks

in the screening context are often at least partly *preventive* in nature; aimed to be undertaken before a person has fallen ill in order to secure (or make more probable) that she does not do this or, if she does, that forthcoming symptoms and consequences are less serious. This fact actualises the alleged *prevention paradox*, according to which most screening (and other preventive) programmes that are effective at a population level (visible through effects on overall morbidity and mortality rates) will necessarily bring very little (if any) health benefit to (most) single individuals.[4] However, this appears to be paradoxical only if a number of assumptions are made.

One such assumption is that effects that are merely visible as probabilities cannot constitute benefits, since the undertaking of medical investigations and tests will affect the risk of contracting a disease run by particular individuals, thus meaning that those who are found to have no (or very low) risk can be identified as such (instead of being seen as running the initial risk assignable to any random member of the population on the basis of the statistical incidence of the disease at the population level).[5] This point could be further expanded into the idea that if such adjustment of the assessed risk is a benefit, then the *opportunity* of having it is also a benefit. Another assumption is that there is something incoherent with the idea of running a programme where the obvious benefits are most visible on a population level, but where the random individual will enjoy a very slight statistical "share" of that interest. However, this assumption, just as the former one, has nothing to do with logic or semantics, but is in fact a substantive and controversial ethical conjecture. It does point to the sort of (potential) tension between the public health ethical and the health care ethical perspectives on screening indicated earlier,[6] but this tension is not in itself a contradiction, but rather a topic for analysis and discussion. We will be revisiting it several times throughout this book.

However, even if all of this is acknowledged, it is far from clear which treatments should be judged as *acceptable* and for what reasons. For some diseases, for instance malaria or polio, there are efficient preventive measures with few side effects, while other treatments are more arduous (like surgical removal of tissue or organs in order to prevent cancer[7]) and/or merely delay certain symptoms (like treatments for HIV or Alzheimer's

---

[4] Rose (1992), p. 12.

[5] Verweij (2000), pp. 51–52.

[6] Cf. Verweij (2000), pp. 52–53.

[7] For example prophylactic mastectomy (removal of breast tissue), hysterectomy (removal of the uterus), and colostomy (removal of (parts of) the colon).

disease). Regarding genetic disease, PKU is one of very few serious monogenetic diseases that can be efficiently and easily prevented (by restricting the intake of phenylalanine in diet), but also this treatment involves rather burdensome life-style adjustments.[8] Most other monogenetic diseases cannot be treated or ameliorated at all, although the assessed risk of contracting such diseases may be clarified by having a genetic test. However, some symptoms of some chromosomal and monogenetic disease can be ameliorated to some extent, e.g. cardiac malformations connected to Down syndrome, and dysfunctions related to cystic fibrosis. Not only do these examples demonstrate that discussions about screening programmes must take the development of new treatments into account. They also show that there is no clear-cut answer to the question of what treatment should be deemed as "acceptable enough" to warrant screening. This also holds with regard to the idea that having oneself tested in the case where one's risk turns out to be low or non-existent provides a benefit in the form of a reduction of assessed risk. Marcel Verweij has remarked that even if the opportunity of having one's estimated risks of disease thus reduced is a benefit, it does not follow that the measure available to effect this risk reduction provides a *net* benefit, and has extended this point to the case where risk-reduction is available also for those who are found to run a high risk of future ill-health.[9] We will return many times to the various ways in which the offer of screening, the undertaking of some testing or the provision of the information produced by that testing may burden the individual person.

As a working hypothesis, we will adopt a flexible view on the matter of when a proposed screening programme may be said to involve or constitute an acceptable treatment, according to which more serious symptoms may make an otherwise too burdensome or uncertain treatment acceptable. However, as will transpire later on, the matter is even more complicated and where the line should be drawn in particular cases remains a difficult issue even if this general line of thought is accepted.

Moreover, it is unclear what should be included in the *concept* of treatment to start with. Many of the examples mentioned above focus on treatment in the traditional sense of *medical* treatment, i.e. procedures manipulating biomedical factors with the aim of preventing, curing or ameliorating (symptoms of) some disease. However, we have also seen that the testing, the offering of testing, or the information coming out of a test, may

---

[8] Many of the diseases targeted in typical neonatal screening programs share this feature (see 3.1.2 below).

[9] Verweij (2000), p. 56.

2.1 Screening, Treatment and Prevention: Preliminary Remarks

be seen as a treatment in its own right (since it may provide the benefit of reducing one's initial assessment of the risk of ill-health). In line with this, it has been suggested that *counselling* can be seen as a form of treatment also in those cases where the test-result is positive, for example, in connection to prenatal diagnosis and genetic testing.[10] The rationale behind this suggestion is complex, one part being about the perceived need to inform patients about environmental and life-style factors open to adjustment through the choices of patients. Another part instead connects to the inclusion of other goals in health care and medicine than the traditional ones of combating disease, an expansion that has, in the least decades, been most conspicuous in areas such as reproductive and genetic medicine,[11] but lately has come to be proposed as valid for health care in general under the heading of patient- or person-centred care.[12] These additional goals are almost without exception cast in terms of either some kind of *psychological well-being* – e.g., the reduction of anxiety, the deliverance of recognition and reassurance, the facilitation of preparation – or in terms of the promotion of *autonomy*.[13]

This last value is often invoked when discussing prenatal diagnosis and related genetic services. In the same context, the possibility of preventing the birth of children with some disease is also invoked as a rationale for screening, thus holding out abortion as a treatment in the screening context. This last point is, of course, highly controversial,[14] but so is the idea of the patient's personal life-plan determining what is ethically defensible to do in the health care context.[15] We will return to these issues in Chapter 3.

Although the advantage of calling measures that further these kinds of values "treatment" is unclear, they can nonetheless be regarded as serving goals that may justify the implementation of some screening programme.[16] Thus, somewhat roughly, there are three kinds of goals or values that can provide the rationale for introducing screening programmes:

---

[10] Shickle (1999), p. 10.

[11] Juth (2005), chapter 2.

[12] For overviews, relevant distinctions and further references, see, e.g., Mead and Bower (2000); Sandman and Munthe (2010); and Munthe et al. (2011).

[13] Juth (2005), chapter 2; The Nuffield Council (1993); and Sandman and Munthe (2009).

[14] See, e.g., Munthe (1996); and Parens and Asch (2000).

[15] Thus, the idea of letting patient autonomy be the primarily guiding principle for what may be done within health care has been criticised for letting medical ethics give way to a purely consumerist model of health care (Eddy, 1990), which in turn has been held out as a threat to public health (Munthe et al., 2011).

[16] We will not enter any debate on the terminological question of whether or not these values should be subsumed under the term "treatment" or not, since it is irrelevant to the question of what screening programs are justified.

1. Improvement of (physiological) health, reduction of disease or amelioration of symptoms of disease.
2. Improvement of psychological well-being or reduction of suffering.
3. Promotion of autonomy.

An important basic ethical issue is, of course, to *what extent* these suggested goals really are worth promoting. Not least since the basic values underlying these goals may conflict in various ways.

Related to basic values usually invoked in health care and medical ethics, points 1 and 2 regard different aspects of well-being and point 3 regards autonomy. The difference between points 1 and 2 is that point 1 regards the traditional goal of preventing, curing or reducing physiological (symptoms of) disease, while point 2 regards the attitudes, feelings and thoughts resulting from having certain information about one's risk of disease, e.g. reduction of anxiety or uncertainty. However, both these goals ultimately aim at improving the quality of life of some individual(s). Improving the quality of life of someone can be done in two ways: by reducing "ill-being" (e.g. by reducing anxiety or ameliorating symptoms of a disease) or by improving well-being (e.g. by reassurance that risk of disease is low or by prolonging a life worth living). However, since well-being includes different aspects, there may be conflicts of well-being. For instance, learning that one has an increased risk of developing diabetes can lead to increased life expectancy due to adjustments of life-style while at the same time increasing anxiety. A number of other potential value conflicts will be described in Sections 2.3–2.4.

2.2 Health: Life and Well-Being

As previously pointed out, well-being can be increased (or "ill-being" reduced) by screening in mainly two ways: by increasing physiological health, which is the most uncontroversial rationale for screening, or by increasing psychological well-being.[17] The most straightforward way in which an individual may be better off regarding her psychological well-being is by having her anxiety or uncertainty regarding her future health status removed – e.g. by having an initial assessment of risk of disease adjusted by having some test. However, the appearance of this possible benefit presupposes that the individual in question has been suspecting that she

---

[17] Psychological well-being in this context refers to all kind of well-being not directly a part of the symptoms of disease or bodily damage.

(or her offspring) is at risk for the disease(s) screened for in the first place (otherwise, there is no initial risk assessment to adjust). Since, by definition, in the case of screening, the initiative to having the test does not come from the individual herself, this cannot be presupposed here. We do not question that reducing anxiety or worry is a valid objective of health care, but this objective is primarily an argument for health care to offer medical investigations *if requested by individuals*, not for screening. On the contrary, screening runs the evident risk of creating more anxiety than it reduces, since, as a rule in screening, the individual has no prior suspicion of being at risk. Being offered or advised to undergo an investigation may therefore give rise to the impression that there is something to worry about.

Thus, in terms of well-being, *screening* programmes (in contrast to the offering of testing to concerned individuals approaching health care with some initial worry) must primarily be justified with reference to their potency for resulting in some kind of improvement of physiological health. Furthermore, there must be reason to believe that organising the offer of a certain testing procedure in the form of a screening has some advantage to the standard model of letting the individual herself take the initiative to health care. So, in order for screening to be justified with reference to well-being, there must be some sort of *treatment* (of the condition tested for in the programme), an *advantage of early detection in terms of this treatment,* and reasons to believe that *the individuals themselves have no prior suspicion of being at risk*. Since, as noted in the foregoing section, each of these factors may be present to different degrees, so it may still be quite difficult to determine whether a particular screening programme is defensible or not. We will return to these more subtle issues in Chapter 3.

2.2.1 Health and Counselling

One important instrument for having a positive outcome of a screening programme in terms of health even if the just mentioned conditions are met is the presence of an appropriate counselling organisation. The role of such an organisation in this context is to secure both that individuals with positive test-results are aware of the presumed beneficial treatment options and how to access these, and that those with negative results are aware of the significance of this information. In cases where the screening targets conditions to which people are continuously at risk (such as dental health screening of children and adolescents), it may also be of importance for the counselling to support those individuals who are found not to suffer from the condition to sustain that state in the future.

The importance of the first task of a counselling organisation should be obvious. If people who enter a screening programme and test positive are unable to use that information for promoting their health, in spite of the presence of beneficial treatments, the programme loses its basic point. A well-known example of what may result if the second part of the task of a counselling organisation fails is the 1970s case of genetic screening for sickle cell anaemia among people of African origin in the U.S. Several people who where identified as healthy carriers of the genetic mutation in question, went on to believe that they were in fact seriously ill.[18] The third task is of particular importance in screening programmes where the main rationale is to secure a certain level of population health, since if healthy people are not supported to stay healthy, the preventive function of such programmes may soon be lost.

Of course, shortcomings of the kind just indicated may be found not only within screening programmes, but also in connection to the offering of checkups and testing within health care in general. However, as noted above, since screening programmes tend to target large and unprepared populations, setting up organisations for appropriate counselling is particularly complicated and challenging and usually requires substantial resources. Thus, the room for serious mistakes is larger, and meeting this challenge makes it more difficult for a screening programme to justify its costs compared to less proactive ways of organising the offering of health care services. This theme is further pondered in Section 4.2.

2.2.2 The Good of People and of the Population

As mentioned, the history of screening is entangled with the history of public health and the aim of promoting overall population health. In many cases of screening, this aim harmonises well with the classic goal of health care and medicine of promoting and protecting the health of individual people.[19] There are, however, specific areas of public health where tensions may appear between these two aspects of promoting health.

One such area is the management and control of communicable disease. Above, HIV-testing of the known sexual contacts of an infected individual initiated by authorities responsible for communicable disease management

---

[18] Hampton et al. (1974).

[19] As mentioned in Section 2.1, this is compatible with a screening programme actually *not* providing a *net* benefit to many individuals.

2.2 Health: Life and Well-Being

was mentioned as an example of medical testing programmes residing somewhere in between pure screening and individual testing. In that case, in affluent countries, the testing has a clear individual health purpose, besides protection against an epidemic. However, in less affluent countries this type of testing initiative from the authorities will many times not be followed by offers of treatment (since there is no money for that), but is only motivated by the public health need of protecting the uninfected part of the population against those infected.

This type of rationale can be envisioned in hypothetical scenarios of more pure screening programmes also in more affluent countries. Imagine, for example, that the a new virus appears that gives rise to a highly communicable and very aggressive and dangerous disease for humans. Imagine further that it turns out to be very difficult to combat this disease with ordinary medical means. In such an imagined case, what remains to be done is to concentrate on traditional public health measures aimed at isolating infected people from the uninfected. Under such conditions, it is not implausible to expect that a general screening programme to this effect may be proposed – indeed, it is highly likely that the programme will even be proposed as obligatory. In fact, recently, it has been suggested that, in light of the global burden of infectious disease, reasons of the sort just cited may support obligatory screening programmes targeting, e.g., international travellers already in the present situation.[20]

Albeit hypothetical, such scenarios expose clearly the potential tension between the traditional goal within health care of promoting individual health, and the goal of public health to promote the health of the population. It also illustrates how the rationale of screening programmes may oscillate between these two goals, and how such a movement with regard to rationale changes what may appear as ethically defensible.

In the case of threats of serious and major pandemics of the sort just envisioned, many people would presumably be willing to accept that the health and autonomy of some individuals is sacrificed for the sake of the long-term health of the population. However, there are other, structurally similar, cases where the spontaneous opinions of most people bend in the opposite direction. For example, below (Section 3.1.1) we will address the idea that the point of prenatal screening is to reduce the number of disabled people born in society. Most people seem to be disturbed by the suggestion that prenatal diagnosis in this way mainly aims at benefiting society as a whole (through decreased expenditures for the care and support of the disabled).

---

[20] Battin et al. (2008).

And very few would be willing to even play with the thought of having such considerations justify infringements of individual liberty.

2.3 Autonomy

Autonomy – or personal autonomy – is a notion that relates to how well a person is able to control her life according to her own plans through her own decisions and actions. It is customary to distinguish between, on the one hand, autonomy as a feature of a *person* (in which case this person's capacities for forming plans, making decisions and executing these are in focus), autonomy as a feature of a person's *life* (in which case, the notion pinpoints the degree to which an autonomous person succeeds in controlling her life with the help of her capacities), and different *ethical ideals* that make use of these two notions. Such ideals specify to what extent and why we have moral reasons, e.g. not to interfere with autonomous people who try to exercise control over their own lives.[21]

The idea of having the promotion of (health related) autonomy as a goal of health care in general and screening in particular is such an ethical ideal, but it is both of later date and more controversial than is the goal of improving health. The idea of having autonomy as a goal of health care has primarily been proposed in relation to reproductive medicine, originating in particular within the practice of prenatal diagnosis – thus specifically focused on the improvement of reproductive autonomy by increasing couples' and women's reproductive options. However, the idea of improving autonomy need not be confined to the reproductive area. Of particular relevance to screening is the fact that the promotion of autonomy has recently received increased attention in discussions and policies with regard to general public health.[22] In addition, in general health care ethics, the above-mentioned trend towards so-called patient- or person centred care holds out the promotion of autonomy as a central objective for all of health care.[23]

The line of reasoning underlying the idea that screening improves autonomy is roughly the following: if individuals possess the knowledge that they or their offspring have an increased risk of contracting some disease, they are in a better position to plan their lives in accordance with their own conception of a good life, to live in accordance with their own values or basic

---

[21] See Juth (2005), chapter 3.
[22] Munthe (2008).
[23] Sandman and Munthe (2009).

wishes or to realise their own important projects (or something like that). To live such a life is roughly what is means to live an autonomous life or being an autonomous person, according to traditional general accounts of autonomy.[24] The idea is, then, that leading an autonomous life, or at least increasing the possibility of doing so, is something that screening may promote.

2.3.1 Respecting and Promoting Autonomy

Traditionally in ethics, autonomy has not been considered primarily as a value to promote, but rather as something that gives rise to moral restrictions on how we are allowed to treat each other when trying to promote other things found valuable. The idea is that if an individual is adult and competent to make decisions, other individuals should not prevent that individual from making decisions and act upon them, as long as she does not violate the rights of others[25] or inflict harm in a wider sense on someone else.[26] According to this line of reasoning, we thus have a strong moral reason not to restrict the autonomy of others. This, however, is not normally thought to imply any obligation to help others to become *more* autonomous or live more autonomous lives. In other words, according to this line of thought, there is a moral obligation to *respect* autonomy but not necessarily to *promote* it.

This latter way of thinking about autonomy is also the predominant one in medical and health care ethics.[27] It is the standard argument against manipulating or coercing people into undergoing medical procedures, such as tests or treatments. Instead, the patient is held to have a right to know what the measure in question is about and a right to accept it or reject it, i.e. *informed consent* should be obtained from the patient (see Section 4.1). This is well in line with the idea of autonomy as something that ought to be respected. We will return to issues about what this idea may imply for the ethical assessment of various screening programmes in Chapters 4 and 5.

In contrast, the idea presently being considered is the suggestion that autonomy is also conceived of as a value that ought to be *promoted*. The point is thus not only to respect people's wants when trying to promote their

---

[24] See Juth (2005), chapter 3.
[25] Locke (1689); and Nozick (1974).
[26] Mill (1859); and Glover (1977).
[27] Beauchamp and Childress (2001), chapter 3.

health, but also to enable them to become more autonomous. That is, autonomy is not only the foundation of restrictions on how health care is allowed to treat people, expressed in terms of duties and rights, but *a value the promotion of which may provide a rationale for health care procedures*, e.g. screening programmes. At the same time, the idea of autonomy as founding duties and rights is not entirely cast away. On the contrary, notions of rights "to know" or "not to know" are a recurring theme in the ethical discussion of, e.g., genetic screening.[28] So, apparently, the idea is that autonomy can be seen both as a value to promote *and* as a basis for restrictions on the promotion of this as well as other values.

The notion of a right to know suggests the idea that promoting people's autonomy is a good thing. The often-acknowledged right *not* to know implies, however, a rejection of the idea that this value of autonomy may justify a *duty* to know, let alone a right for others to have people perform this duty.[29] However, the rejection of a duty to know can also be interpreted as recognition of the fact that more information about one's health is not necessarily conducive to autonomy.[30] For instance, information about risk of future disease can make an individual depressed or confused. Moreover, such information can make a cautious person cancel plans she would have gone through with, had she only had a go at them. That is, even if the information that may be revealed by participation in a screening programme may be of relevance to one's ability to lead one's life in accordance with one's plans, it does not follow that actual revelation of this information will *in fact* provide help in this respect. For example, the context and manner of *disclosing* the information may have an undue manipulative effect, rather than an autonomy enhancing one. Of course, although medical information need not enhance the autonomy of an individual, it may also do so in many cases. From the point of view of autonomy, the task is then to disclose the information in a way that is "autonomy-enhancing", a task undertaken by counselling services. (see Section 4.2).

Of even more practical consequence, is that the notions of respecting and promoting autonomy are in potential conflict. Just as different aspects of well-being may pull in opposite directions in a particular case, respecting a person's autonomy may in a particular case have the effect that this

---

[28] Hoedemaekers (1999). See also Sections 5.1 and 5.2.

[29] There are, we admit, suggestions to the effect that people do have a duty to know things of importance for their life-plans, such as health risks (Rhodes, 1998). However, such ideas are generally discarded in the area of medical ethics (Beauchamp and Childress, 2001, pp. 61–63).

[30] Juth (2005), pp. 305–309; and Sandman and Munthe (2009).

2.3 Autonomy

person is allowed to make choices that makes her life less autonomous. Accepting the promotion of autonomy as a goal of health care in general, or of screening in particular, thus give rise to the issue of whether or not the importance of respecting autonomy may be overridden in the light of the value of enhancing autonomy.[31]

2.3.2 Promoting and Respecting Autonomy Through Screening

Although the autonomy of many individuals may be promoted by receiving information from medical investigations, there are special obstacles for realising such an end related to screening programmes. There is also a more basic issue of how to interpret an idea of promoting autonomy regarding practices taking place, at least partly, within a public health context. Even if autonomy is accepted as a value, it is difficult to have public health practices apply fully the individualistic notion of personal autonomy traditionally employed in health care ethics. In particular, the strict requirement of *respecting* personal autonomy seems hard to fully reconcile with the population perspective of public health.[32] This gives rise to a complex and general problem for the ethics of screening to which we will return in Chapter 6. The observations below apply standard opinions in medical and health care ethics, only noting in passing how the adoption of a public health perspective may change the picture.

As previously noted, a central characteristic of screening is that the initiative to the investigation is not coming from the individuals themselves, but from society or large societal institutions. This is potentially problematic already from the point of view of the idea of *respecting* autonomy, since that idea makes pressure to accept or abstain from medical procedures at least questionable. Respect for autonomy is usually not only taken to mean that informed consent should be obtained in order for an investigation to be justified, but also that the individual should not be pressured or subjected to more subtle manipulative efforts.[33] At the same time, defensible health care ethical standards may give room for some pressure in the form of persuasion, at least if such an action is transparent, in the relation between a doctor

---

[31] Juth (2005), pp. 100–108.
[32] Munthe (2008).
[33] Beauchamp and Childress (2001), pp. 94–95.

and her patient.[34] However, when screening is on the agenda, the very *institution* of health care (not only your personal physician or GP) is applying the pressure – often as a representative of society as a whole. Hence, it cannot be denied, that some substantial pressure beyond concerned persuasion is being applied on the individual to accept rather than the reject the investigation in the case of screening.

Of course, we are dealing with matters of degree here. The more that participation is taken for granted in the offer of entering a screening programme, the more pressure there is to participate. The more benefits are emphasised and the less drawbacks are mentioned, the more pressure to participate. The more active the patient has to be in order to refuse investigation, the more pressure to participate. And the more pressure to participate, the more problematic from the point of view of autonomy.[35] Although pressure can be reduced in various ways, the general point remains: the very fact that the initiative comes from society always presents some problem from the point of view of autonomy.

In addition to this, screening programmes may have side-effects on the societal level that are detrimental to the autonomy of people.[36] If screening is readily available, it may be looked upon as irresponsible to decline some testing, since screening being generally offered may reinforce the norm that we are responsible for our own health and the health of our children, or, for that matter, of the health of the population. This, in turn, may create a significant social pressure to participate in the screening programmes. For example, this observation is a recurring theme in debates on prenatal screening (see Section 3.1.1 below).

At the same time, from the perspective of public health, promoting autonomy at a *population level* may very well be seen as compatible with having some such side-effects in terms of the reduction of the autonomy of some individuals in some situations. For instance, facing a serious pandemics scenario, an idea of each citizen being provided with equal opportunities to promote their health may imply rather substantial pressures on individuals to test themselves for the disease in question. Similarly, just upholding a level of population health enabling a society to provide good health services may require an outspoken imperative vis á vis people in general to look after and take responsibility for their own health.[37] In other contexts, such tendencies to disregard the autonomy of some individuals in some situations in

---

[34] Sandman and Munthe (2009).
[35] Clarke (1998), p. 401. See also Section 5.1.
[36] Hoedemaekers (1999), p. 212.
[37] Cf. Dawson and Verweij (2007b); and Munthe (2008).

2.4 Justice

the name of public health and autonomy may instead be seen as particularly disturbing from an ethical perspectives, e.g., in the area of prenatal diagnosis and reproductive medicine (see Section 2.2.2). We will elaborate further on these themes in later chapters and discuss more closely what importance that should be attached to individual autonomy in different contexts of screening in Section 6.5.1.

Finally, and this applies also from a public health perspective on autonomy, just as in the case of the goal of promoting health, there are practical problems related to screening regarding the need for counselling in order to promote autonomy. We will return to the case of the need for counsellors in order to preserve and promote autonomy in Section 4.2, but it should be noted that the significance of counselling seems both stronger and more general in the case of promoting autonomy than in the case of promoting health. Again, it is worth repeating that proper counselling takes time and resources. Considering the volumes of people involved in screening programmes and the relative unpreparedness of these, counselling comes out as difficult to combine with the common criterion of "[e]conomic balance of the cost of case-finding in relation to total expenditure on medical health care".[38] There is, in effect, always a counter proportionality between pursuing the promotion of autonomy as a goal and economic viability in the context of screening.

2.4 Justice

Besides the promotion of health and autonomy, another basic value that is sometimes mentioned in the context of screening is *justice*.[39] Even though justice, like autonomy, is a contested concept, there is a basic conception of justice common to all suggestions about what justice is – namely the idea that *relevantly similar cases should be treated similarly*.[40] This type of idea has been used to defend screening on the basis that if some testing procedure is offered due to its benefits for someone, all those who may gain from these benefits should receive the same offer.[41]

Nevertheless, our contention is that justice is secondary to the values of well-being and autonomy in the screening context. This since considerations

---

[38] Wilson and Jungner (1968), p. 35.
[39] Hoedemaekers (1999), p. 209.
[40] See e.g. Rawls (1972), p. 5.
[41] Shickle (1999), p. 22.

of justice work only as long they are about the equal opportunity of people to access what is clearly a *good*. This holds only if one can presuppose that the benefits befalling the initial (limited number of) people by having some type of test would remain a benefit when health care approaches large numbers of people who would otherwise not have contacted health care themselves. However, as has already been indicated and as will be elaborated further in later chapters, there are reasons to be sceptical about that, since the very move from testing on the initiative of individual requests to screening may give rise to a number of side-effects. As a rule, such a supposition can only be made if the benefits are great and obvious enough while the downsides are few and rather mild, e.g. if the disease in question is very serious, if it can be prevented efficiently without too many side-effects, if early detection is an advantage and if there are good reasons to believe that those contacted have no prior suspicion of being at risk. Otherwise the downsides in terms of being worried and pressured may outweigh possible advantages for some individuals, in which case the premise that they will be benefited is not fulfilled.

At the same time, considerations of how to distribute benefits and burdens in a fair manner appear to become central with regard to several issues touched on later in this book. The basic problem of how to trade off false positives and negatives taken up in Section 3.2.2 seems to be about that. The idea of having the promotion of autonomy as a goal of screening also from a public health perspective seems to be most plausible when cast in the form of an idea of the equal distribution of opportunities to improve health[42] – and, as just have been seen, a central issue for the ethics of screening is what importance such an idea is to be accorded. Nevertheless, as will be demonstrated, such issues seem to a large extent to be approachable without *assuming* some specific theory of justice. Rather, the role of considerations regarding justice in the screening context will generally be determined by the outcome of considering the sort of ethical issues already sketched.

For this reason, with one notable exception, well-being and autonomy remain the *primary* candidates for values that may be promoted by screening. Justice becomes relevant primarily as an additional argument in those cases where the other values speak in favour of screening. In effect, we will concentrate on those latter values in our discussion. The one exception to this is Chapter 6, where we will touch on the more overarching issue of how the pursuit of, by themselves ethically acceptable, screening programmes should be ranked relative to other important social undertakings (Section 6.5.3).

---

[42] Munthe (2008).

Due to the elevated costs of such programmes, this general issue regarding the distribution of resources in society seems to us to be the central one from the point of view of justice raised by screening.

2.5 Summary

In order to provide a rationale or justification for any screening programme one has to have some idea about the values that the programme should promote. There are three kinds of values or goals invoked as the rationale for introducing screening programmes: improvement of public health, improvement of psychological well-being or promotion of autonomy. We point at several conflicts within and between these goals, e.g. the potential tension between the traditional goal within health care of promoting individual health and the goal to promote the health of the population. We also argue that screening programmes must primarily be justified with reference to their potency for resulting in some kind of improvement of physiological health, e.g. since screening runs the evident risk of creating more anxiety than it reduces. Moreover, the very fact that the initiative does not come from the individual herself always presents some problem from the point of view of autonomy. Considerations of justice become relevant primarily as an additional argument in those cases where the other values speak in favour of screening.

Chapter 3
Screening – What, When and Whom?

Every screening programme has to be evaluated in terms of the goals it tries to achieve – how well it may be expected to achieve them, what possible drawbacks in terms of counteraction of some goals and other undesirable side-effects it may bring, and the overall balance of its costs and benefits compared to alternative measures. In the previous chapter, the possible general goals of screening and some of the various complications that they imply were introduced. In the present one, we will be addressing in more detail various problems that arise as a result of certain kinds of screening.

Screening programmes may be more or less warranted depending on properties of the disease or health problems targeted by the program, the testing, investigatory and analytical methods applied, and the treatments available. Since different diseases have their onset in different ages, there is also the related question of when in people's lives tests should be made – during the prenatal, neo-natal, childhood or adult stage. For this reason, we will start out in Section 3.1 by addressing the issue of what diseases to screen for in relation to what groups will in effect be targeted by the programme. Having done that, considerations relating to methods for investigation, testing and analysis will be the subject of Section 3.2, while Section 3.3 will be devoted to the issue of what to require of available treatments for targeted conditions in the screening context.

3.1 Diseases and Groups

Wilson and Jungner included in their classic recommendations the requirement that the disease screened for must be an "important" health problem.[1] However, what this is to be taken to mean is far from obvious. Wilson

---

[1] Wilson and Jungner (1968), p. 27.

and Jungner took "important" to mean at least either that the prevalence of disease in a population must be high, or that the disease is "serious" enough for the individual. Since they used PKU as an example of a serious disease with low prevalence, they probably had diseases connected with much suffering and/or very premature death in mind in the latter case. However, since in many cases one and the same condition can vary in terms of the severity of actual symptoms and degree of suffering from case to case and since what will actually occur with regard to a particular patient is often difficult to predict,[2] this criterion allows for a wide range of interpretations, setting the limit for justified screening on different places on a continuum of seriousness.

Although prevalence, unlike seriousness, is easily quantifiable in numerical terms, there is no clear-cut answer to *how* high the prevalence should be in order to count as "high" in this context. Presumably, a public health perspective was tacitly assumed by Wilson and Jungner on this point, implying that what is to be counted as high prevalence has to be decided in each case in relation to how the condition in question affects population health in relation to the public health related goals of a society. So, for instance, if there was a way to screen effectively for risk factors for the common cold, this would regard a disease with a very high prevalence numerically speaking, but nevertheless presumably not that much of a problem from a public health perspective, and therefore not qualifying for being considered as an "important" health problem.[3] In contrast, seasonal influenza (again, if there was a way to screen for risk factors connected to this) may indeed qualify, especially if it affects seriously not only the elderly and very young, but also a broad spectrum of adults and/or adolescents. In other words, the idea of a high prevalence needs to be connected to some notion of the "seriousness" of a disease in order to make the notion of an "important" health problem intelligible. But then the question is raised what is to count as *serious* in this context.

In an attempt at clarification, the following three dimensions of seriousness or severity of disease have been proposed[4]: degree of harm to health,

---

[2] Take for instance Down syndrome or fragile X, where symptoms and degree of symptoms (e.g. degree of mental disability) can vary greatly and where the prospects of predicting the outcome for a particular individual are poor (Connor and Ferguson-Smith, 1997, pp. 118, 136–137).

[3] At least, this would seem to hold for reasonably developed countries, where the general health status of the population is sufficiently strong to prevent a common cold to develop into a more serious health threat.

[4] Post et al. (1992).

age of onset (the earlier, the more severe) and the probability that the disease will develop given the presence of the risk indicator. Although this makes the notion of a serious disease somewhat clearer, it is still an open question for each of these variables to what degree it needs to be fulfilled in order for screening to be warranted.

However, none of this clarifies how seriousness or severity is to be balanced against prevalence with regard to the question of whether or not a certain disease is to be seen as the sort of "important" health problem that may warrant screening (given that further conditions on tests, treatments, counselling, et cetera are met). As will be demonstrated in the following sections, the actual practice of screening seems to oscillate quite freely between giving prevalence the decisive weight and holding out severity in the single case as the primary reason for launching a screening programme. While the latter emphasis is most obvious in the neonatal area, prenatal screening programmes tend to include a number of conditions of a rather mild nature (e.g., sex-chromosome aberrations, such as Turner's or Klinefelter's syndrome). Possibly, this difference may connect to different goals being pursued through screening in these respective areas.

As already argued, one cannot determine how warranted a screening programme is in isolation from its goals. However, since different screening programmes can (and do appear to) have different goals depending on who is the target of the programme this seems to mean that the assessment of the ethics of such a programme needs to proceed from different outsets depending on the target population. For instance, reproductive autonomy is foremost a goal for prenatal screening, while other screening programmes with other targets aim for other goals. Thus, the goals can affect what conditions should be considered serious enough for screening. For this reason, we will in the rest of this section consider particular problems regarding what diseases to screen for, related to different target groups for screening: prenatal, neonatal, children and adolescents, and adult.

3.1.1 Prenatal Screening

Autonomy and then, naturally, reproductive autonomy as a goal for screening is most salient in discussions about prenatal diagnosis[5]: if the prospective parents can gain knowledge about the expected health status of potential

---

[5] Chadwick et al. (1999). The same goes for genetic carrier detection on adults in populations with higher risk for certain genetic disorders, e.g. thalassemia in Cyprus (Ioannou, 1999).

children, they are in a better position to make reproductive decisions in accordance with their own plans – in practice, primarily decisions regarding whether or not to have the pregnancy terminated.

However, even if one grants that reproductive autonomy is an important goal for the practice of prenatal *diagnosis*, it is a highly questionable argument in favour of organising this practice in the form of *screening programmes*. The reason for this is the same one that makes screening programmes problematic in general from the point of view of autonomy: since the initiative to the investigation in question is not coming from the individual, but from the institution of health care or society, there is always some significant pressure on the individual to undergo the investigation. Furthermore, since screening programmes single out some of all the conditions that can be tested, these conditions will inevitably be perceived as especially problematic, a mechanism which tends to reinforce the initial pressure. Moreover, if, as is the case in many countries, this is combined with legislation on abortion that makes this procedure more accessible for pregnant women who have had a positive test result from prenatal diagnosis, this pressure is extended to the choices of these women on whether or not to carry the pregnancy to term.

Rather, consistent observance of the value of reproductive autonomy mainly serves as an argument for the availability of prenatal diagnosis and against the idea of banning prenatal testing for certain conditions while allowing others.[6] For society to draw such a line is certainly to apply salient pressure on individual prospective parents to make certain reproductive choices while avoiding others. Reproductive autonomy also supports the idea of having health care provide every pregnant woman with the same general initial information about available prenatal services, while leaving the decision to look for further information or making use of these services to the women themselves. However, none of these measures amount to *screening* – on the contrary, autonomy alone seems mainly to tell *against* prenatal screening (while supporting prenatal diagnosis organised in other ways). Somewhat metaphorically, autonomy or freedom resides between external coercion and individual obligation and while forbidding or banning some prenatal testing is coercion (not to do it), screening draws towards creating an obligation (to do it), psychologically and/or as a part of the enforcement of social policy.

Even though reproductive autonomy is nowadays a very commonly mentioned rationale for prenatal screening, the promotion of health and

---

[6] See Buchanan et al. (2000) for discussions regarding this question and further references.

3.1 Diseases and Groups

prevention of disease has been an important historical argument for prenatal screening programmes.[7] However, since very few of the conditions targeted by prenatal screening can be prevented and since prophylactic measures to delay and reduce symptoms are usually not dependent on *prenatal* detection,[8] prevention of the actual disease that the individual may develop can seldom be a reason for *prenatal* screening. Rather, what can be avoided is *the very existence of a future individual that would, if born, (probably) develop some disease*. In effect, the follow-up procedure offered in almost all cases of a positive test result in prenatal diagnosis is the opportunity to stop the further development of the foetus by termination of pregnancy, i.e. abortion. To this may be added that some of the testing methods used for prenatal diagnosis (amniocentesis and chorionic villus sampling) pose significant health risks regardless of the health status of the foetus. We will return to that issue in Section 3.2.

The fact that the only way to "prevent" most conditions that can be detected by prenatal diagnosis is abortion makes this procedure ethically controversial. Even the very use of the notion of prevention has been held out as suspect from a medical and health care ethics perspective, due to the apparent difference between preventing the occurrence of disease in a person and preventing the occurrence of this person in the first place.[9]

To this may be added that, if the rationale of prenatal screening is to be the prevention of disease, the most common prenatal screening programmes today clearly illustrate the above-mentioned general tension between a health care ethical and a public health ethical perspective on screening, as well as the connected issue of how to balance severity and prevalence. These programmes are as a rule not primarily targeting the most serious conditions that may be revealed by prenatal diagnosis (such as Tay-Sachs or Werdnig-Hoffman disease), but mainly moderately serious conditions, such as Down syndrome, and several milder chromosomal conditions. In the last decade, this has been made particularly salient by the selection of chromosomes to be investigated necessitated by the new QF-PCR method,[10] where prevalence

---

[7] Munthe (1996).

[8] One of few examples is the surgical procedure of twin-twin transfusion, where the shared vessels of twins who have the disorder are separated by laser (personal information from The-Hung Bui, professor of clinic genetics at the Karolinska University Hospital, Stockholm).

[9] Hildt (1999); and Munthe (1996), chapter 3.

[10] This method is more rapid than the traditional charyotyping method, but while the latter allows investigation of all chromoses, the former allows only the investigation of a selected few ones – in its standard version, the chromosomes 13, 18, 21, X and Y. See, e.g., Vahab Saadi et al. (2010).

seems to be the main aspect considered as relevant in the application of this method. This fact has, in turn, provoked some debate also within medicine, since some very serious, albeit rare, conditions will remain undetected.[11] If the goal of the programme is the prevention of disease (as it would seem to have to be considering the weak case for prenatal *screening* as a means to the promotion of autonomy), apparently, the conditions targetted by QF-PCR have been selected due to their prevalence in combination with an evaluation of the importance of this prevalence from a public health perspective, e.g., regarding the societal costs for care and support of children with such conditions. This picture of what prenatal screening seems to be about can, as will soon be demonstrated, be fed into recent criticism of this practice.

Traditionally, prenatal diagnosis has been accused of being morally dubious due to moral resistance against abortion in general, often based on some principle of the sanctity of life. However, this kind of resistance has lost ground, especially in practical policy making. A general legal right of women to decide on their own whether or not to continue a pregnancy is accepted and implemented within certain limits with few exceptions throughout the developed world. In particular, this right tends to be safeguarded in the context of prenatal diagnosis – also by parties otherwise morally opposed to abortion.[12] Moreover, an argument from the morality of abortion in general does not really target prenatal diagnosis; it attacks the morality of a procedure that may or may not follow a prenatal test. Since most such tests secure a *negative* diagnosis (thus, in practice, motivating the avoidance of abortion), it is, however, questionable to what extent a negative moral view on abortion can be extended to prenatal diagnosis as such. Even less can this type of argument tell against prenatal screening, since this is merely way of *organising* the practice of prenatal diagnosis.

Another source of criticism is instead the idea that prenatal testing for diseases and disabilities is incompatible with ascribing the same value to all human beings, and that this holds especially in the case of screening programmes.[13] The background of this criticism is the fact that prenatal screening is always targeting a specified "list" of conditions and that abortion is the expected response if any of these conditions are identified in a foetus. This has given rise to the suspicion that the goal of prenatal screening is to "weed out" certain types of individuals who, due to their disabilities,

---

[11] Bui and Nordenskjöld (2002).
[12] See, e.g., Munthe (1996), p. 46, n. 1.
[13] Parens and Asch (2000).

3.1 Diseases and Groups

are not seen as worthy of protection to the same extent as are other human beings – possibly due to they being costly to society. At least, it has been claimed, prenatal testing in general and prenatal screening in particular *expresses* such a point of view through its practice.[14]

This criticism of prenatal screening can be interpreted in (at least) two main directions. First, it could be understood as a moral criticism of individual women's or couple's intentions, purposes, or motives to accept the offer of screening in order to abort if the findings are positive. This is, then, not really an argument against abortion as such, since abortion can be performed also without screening. Rather, the argument is that it is especially problematic from a moral point of view to be prepared to "select" one's offspring on the basis of some of its potential characteristics.

However, this objection rests on a rather malicious and misleading portrait of the general motives of those who choose prenatal diagnosis. A woman that chooses abortion without any kind of prenatal diagnosis may do this partly on the basis of general worries that the child may be disabled in some way. Moreover, if this interpretation of the motives behind choosing prenatal diagnosis is to be a part of a consistently applied interpretative scheme, we would have to say that a woman who chooses abortion due to social factors, such as abandonment by the father or destitution, is thereby unfolding a devaluating view of children living in such social circumstances (being fatherless or being brought up in very poor surroundings). However, a more plausible picture is that, in all these cases, including the cases where abortion is the outcome of prenatal testing, the motive could be described as avoiding a situation for the woman or couple in question that she or they consider as unbearable. In addition to this, it is not always the case that abortion is the only measure that potential parents consider: some prenatal testing may be accepted only in order to prepare oneself for the birth of a child that may have special needs.[15] Furthermore, if autonomy is assigned any weight in health care, questioning the motives of individuals accepting measures that health care offers is certainly problematic from a moral point of view.

Second, the criticism can be interpreted as being directed towards the institutions of prenatal testing rather than the individuals who choose to utilise the services provided by these institutions. The idea is then that the offer of prenatal testing in conjunction with the possibility of abortion

---

[14] Munthe (1996); and Parens and Asch (2000).

[15] This seems to be an extremely plausible hypothesis in the case of routine obstetric ultrasound (the most widespread practice of prenatal screening), since this type of investigation reveals a number of factors of direct relevance for the safety of the pregnancy, delivery and the health of the child.

expresses the view that individuals who have the conditions tested for are of less worth than others. This criticism is sometimes rejected with reference to the argument that devaluing some condition does not imply devaluing the individuals who has the condition in question.[16] However, the relevance of this response can be questioned in the prenatal context, since abortion means that the individuals with the condition are discarded and not just the condition.[17] If we add to this the element of societal pressure involved in *screening*, it is hard to escape the impression of the rather large apparatus set up by society to accomplish prenatal screening that avoiding such individuals is a strong wish of society.

However, also this argument can be interpreted in two separate ways, depending on how the idea that prenatal screening "expresses" a devaluing view of disabled individuals is understood. Sometimes the argument seems to say that implementing prenatal screening programmes will strengthen already existing negative attitudes towards disabled people. This, in turn, is predicted to lead to increased discrimination and stigmatisation of these individuals. It has also been claimed that less resources will be spent on disabled individuals, since it is conjectured that these will be fewer and thus weaker as a group if efficient prenatal screening programmes are implemented.[18] Concern has even been expressed that implementing such programmes is to start walking down the path ultimately leading to the resurrection of Nazi-style "euthanasia-programmes" for disabled people.[19]

Interpreted in this direction, the argument regards *the actual consequences* of prenatal screening, i.e., it is founded on empirical hypotheses. However, there is no empirical evidence in support of the position that more prenatal testing leads to more discrimination and stigmatisation of the disabled.[20] Rather, the trend in many countries seems to be the opposite: in tandem with the development of prenatal diagnosis, more resources have been invested in improving both the quality of life and the societal opportunities of disabled people.

However, the argument can instead be interpreted as saying that the explicit or implicit purpose or goal of prenatal screening is to avoid the existence of certain types of individuals due to overarching societal considerations. Or, at the very least, *the institution and actual practice of prenatal*

---

[16] Buchanan et al. (2000), p. 280.
[17] Scott (2005), p. 72.
[18] Scott (2005), p. 70.
[19] Parens and Asch (2000). For a general discussion of these kinds of "slippery slope" -arguments in medical ethics, see Beauchamp and Childress (2001), pp. 144–146.
[20] Buchanan et al. (2000), p. 266, note 3.

3.1 Diseases and Groups

diagnosis clearly communicates the message of such a purpose to the general public.[21] There certainly is something to this criticism, especially when directed towards prenatal *screening* and not prenatal diagnosis in general. If health care is to implement prenatal screening, it has to choose some of all the conditions that can be tested for.[22] This will mean at least some extra societal pressure to test for these things and the primary goal of such a practice cannot, therefore, be to promote autonomy (since such a goal would rather suggest an organisation built on the initiative of individual parties interested in prenatal diagnosis, see Section 2.3 in the preceding chapter). Rather, the goal of preventing the existence of certain types of people seems to be the least farfetched interpretation of the rationale of such screening programmes. What is more, since prevalence in combination with a public health perspective in terms of societal costs for care and support on how to assess the importance of this prevalence seems to be what is actually applied when selecting the target conditions, the message conveyed about what is problematic with having these individuals around is about them being a burden on others and society. However, if this is the goal communicated and health care, through selecting some conditions for screening, thus provides an explicit list of which conditions that are especially desirable to avoid on these grounds, it is hard not to see this as an official ideological message that the individuals having these conditions are considered as being less worthy of protection. At least, it should come as no surprise if those who live in society with the conditions in question forms such an impression. The general acceptance and promotion of the view thus communicated has, moreover, been held out as a risk factor for prenatal diagnostic practices actually taking that eugenic turn emphasised by some critics.[23] This analysis seems particularly warranted in countries with weak systems of care and support for the sick and disabled, and/or bad access for these groups to various other goods readily available to most citizens.

Moreover, this official message thus expressed by prenatal screening programmes feeds back also to the initial pressure on individual pregnant women and couples, thus reinforcing the scenario of a eugenic turn. The

---

[21] Asch (2002); and Parens and Asch (2000).

[22] This aspect of prenatal screening has been accentuated by the recent and earlier mentioned adoption in many countries of the so-called QF-PCR method for prenatal chromosome analysis, where only some selected chromosomes are investigated, rather than the whole karyotype. This necessitates a choice of *which* chromosomes to analyse, and whatever choice is made, it may be held that the implied societal message about the undesirability of certain kinds of people will be more salient.

[23] Munthe (1999), chapter 7; and Munthe (2007).

initial pressure of the suggestive societal offer is added to by the implied message of society that it would rather not have them have certain types of children. Especially so, if also the abortion legislation is suggestive in a similar way, making abortion more easily accessible if a positive diagnosis has resulted from the participation in a prenatal screening programme. Again, deficient support and equality for disabled people adds to this mechanism. In both these ways (devaluating societal messages regarding disabled people and reduction of reproductive autonomy), then, prenatal *screening* – rather than prenatal testing – seems to be especially problematic from a moral point of view, at least in the absence of treatments other than abortion.

This situation has some serious implications for the way in which, in many countries,[24] obstetric ultrasound and, more recently, new methods for safe and more precise risk assessment regarding the presence of chromosomal aberrations in the foetus[25] have been organised as screening programmes, rather than as a routine offer to those who seek invasive prenatal diagnosis on their own initiative. In view of the obvious risk of such an organisation of the practice to express discriminatory messages about disabled people, to undercut patient autonomy, and noting that the chief aim of these methods (to reduce the amount of invasive tests) can be attained without organising the practice as a screening, it is difficult not to suspect that other forces rather than consequent medical ethics and rational argument has been influencing the chain of events. In particular, the attraction for a hospital of being in charge of a large screening programme in economic and other terms, mentioned at the outset of this book, comes to mind. Another factor that must be considered is the interests of commercial medical technological companies who provide the testing-kits for the new risk-assessment methods. From *society's* point of view, however, it would rather seem that avoiding the screening solution is preferable in terms of both efficiency and efficacy.

At this point, it may be asked if these arguments really make up a conclusive case against prenatal screening in all circumstances. The combined arguments rely partly on an analogy between prenatal screening (as opposed to testing available on request) and (unnecessarily) eugenic policies or side-effects and the assumption of such ingredients being seriously problematic. However, it may be asked if not *some* versions of eugenic thinking may in fact be warranted (and thus not unnecessary), at least in some circumstances where it may appear acceptable for a society to temporarily assume a less liberal posture towards its citizens. In particular, may not the objective of

---

[24] E.g. Germany and France (Schmitz et al., 2009).

[25] More is said about these methods in Section 3.2.1 and Chapters 4 and 5.

3.1 Diseases and Groups

decreasing the statistical incidence of diseases that would otherwise require substantial societal resources and severely halt social progress in other areas be an acceptable policy aim *as a transitionary means* for developing countries struggling to build up a modern welfare society which, in the long run, will be of enormous benefit to the entire population? After all, this seems to be what present *developed* countries did in one or the other way back in the days. On a basic ethical level, this is a question about whether or not the tension mentioned in Chapters 1 and 2 between the individualist perspective of traditional health care ethics and the public health ethical perspective may not in this case come out to the benefit of the latter. We will not argue that the answer to this question cannot be affirmative, but we want to end this section by stressing some rather important empirical and practical concerns that have to be thoroughly addressed by anyone tempted to argue along the lines just sketched.

First of all, for such an argument to work it must hold that the society under consideration is in fact in a position where the absence of a prenatal screening programme would actually *be* an obstacle for the sort of generally beneficial social development described. For this to hold, it is not enough that the task of taking care of sick and disabled people is an undertaking seriously burdening the national economy (although that is, of course, a necessary condition). It must also be the case that there are no alternative, less drastic, measures by which the economic hurdles on the path towards the envisioned social progress can be removed or sufficiently reduced. One such alternative may – as pointed out – be to abstain from *screening*, but nevertheless offer the option of prenatal diagnosis in a universal health care system. Moreover, it has to be considered to what extent a society at this level of development would actually be apt at running a prenatal screening programme in an effective way. For example, one reason to consider screening over a general offer of prenatal services including testing on request, may be that the level of education makes it difficult to inform the population of this offer and its significance. However, in that case, it may also be doubted if this society can muster the competence needed for upholding sufficient quality of a screening programme. Furthermore, as noted earlier, there is a connection between the strength of the discriminatory message about sick and disabled people communicated by a prenatal screening programme and the extent to which society makes efforts to support and compensate such people. In the case of developing countries presently considered, we may assume such efforts to be rather modest.

Last of all, the argument at best supports prenatal screening as a *transitionary* measure – when social progress is under way, the reasons for screening rather than the offering of a service on request will quickly disappear. However, some attention here needs to be payed to the apparent fact that the application of those autonomy restrictive reproductive policies (such as compulsory sterilisation) that may be argued to have played a role in the social progress of presently developed societies during the twentieth century tended to linger on in these societies way beyond the point where it makes any sense to ascribe them such a significance.[26] In other words, the idea of installing repressive policies for the sake of the general good must always be assessed not only with regard to their alleged importance and proportionality, but also to the very real risk that – due to socio-political causes – they will in fact stay on well after they have lost their significance.

3.1.2 Neonatal Screening

In some respects, there are similarities between prenatal and neonatal screening. The rapid progress of medical testing methods in general and genetic testing methods in particular have given rise to controversies about which conditions that should be included in screening programmes within the prenatal and neonatal areas respectively. Both prenatal and neonatal screening programmes, moreover, have good chances of being efficient in terms of uptake, i.e. testing a large part of the populations of foetuses and newborns respectively. This is so, since most (prospective) parents are in contact with health care during pregnancy and birth anyway. This also means that both prenatal and neonatal screening programmes are logistically less difficult to organise: there is no need to implement special organisations searching out, approaching and recruiting people for screening. Of course, this is cost reducing and probably one contributing explanation of the fact that all developed countries have more or less ambitious pre- and neonatal screening programmes.

However, there are also many important differences between pre- and neonatal screening. The most obvious one from an ethics perspective is that *reproductive autonomy* is not a goal of neonatal screening.[27] As a result, the question about the goals and thus the criteria for neonatal screening is not

---

[26] For example, in Sweden, while it may be argued that compulsory sterilisation might have had some role in the development of the welfare state (although it is, of course, very difficult to prove that conjecture in retrospect) during the 1930s and 1940s (possibly also part of the 1950s), the policy making room for those measures was not dismantled until the middle of the 1970s. See Broberg and Roll-Hansen (2005).

[27] It has, however, been argued that it could be a goal for future neonatal programs (see Section 5.2). However, no existing neonatal screening program has reproductive autonomy as a explicit goal.

nearly as controversial as regarding prenatal screening. Neonatal screening almost only has been seriously considered for conditions for which there are efficient preventive, curative or ameliorating medical measures – a fact reflecting a broad agreement on the basic goals of such programmes. It is hard – if not impossible – to find documents regulating or discussing neonatal screening where efficient medical treatment of some sort is not mentioned as a necessary prerequisite for a condition to be a candidate for such screening, and when exceptions are considered these are motivated by arguments connecting to the health of the child.[28] Traditionally, then, the goal of promoting health by preventing, curing or ameliorating disease is considered as paramount in neonatal screening.

The most widespread and classic target of neonatal screening is phenylketonuria (PKU). Today, most developed countries have screening programmes for this disease at the national or state level.[29] Many of these countries also have neonatal screening programmes for other diseases, such as congenital hypothyroidism and biotinidase deficiency. These diseases all share some common characteristics: they can be detected at this early age, the condition most likely had not been detected if not screened for[30] and early measures are available as well as necessary for preventing or ameliorating the diseases in question. Moreover, the tests for the diseases are very reliable and safe, and the diseases themselves are very serious. For instance, all the three diseases mentioned leads to severe mental retardation if not treated properly. The obvious safety of the testing methods (common blood samples) is another difference to prenatal testing, where risky invasive methods often are necessary to detect conditions at the prenatal stage (see Section 3.2).

3.1.2.1 Reasons for Screening in the Neonatal Period

The most straightforward rationale for having a screening programme target *newborns* is the presence of benefits due to the very fact that testing is

---

[28] See e.g. Wilcken et al. (2003), p. 2305; Chadwick et al. (1999), pp. 47, 96; and ACMG (2005), p. 5. Some countries do screen for some untreatable diseases, though (Loeber et al., 1999), and a particular fairly recent consideration has been the idea of screening neonatally for Duchenne Muscular Dystrophy (Mohamed et al., 2000). As noted by Nijsingh (2007), however, these are rare exceptions to an otherwise widely observed rule of thumb.

[29] See Wilcken et al. (2003); and ACMG (2005).

[30] The explanation for this is that the diseases are all very rare and, although hereditary, the genetics is not straightforward enough to warrant any suspicion of parents or health care that any particular child is at risk.

performed at such an early age. And there are obvious such benefits if the conditions screened for have the above-mentioned characteristics: diseases that undetected and untreated at an early stage would result in significantly decreased longevity and quality of life for some individuals can then be detected and treated by relatively efficient and simple means. Although each of these diseases is very rare, the facts that so much is at stake for the individual and that the risks associated with testing are virtually non-existing, general screening may thus be considered as warranted.[31]

Neonatal screening for these types of conditions can also be justified with reference to the *best interest* of the patient,[32] i.e. the baby. This kind of rationale is often involved when the patient herself cannot request or consent to the medical measures in question, the potential benefits are great and the risks are slight. Thus, not only are neonatal screening programmes clearly focusing on health as an objective, the notion of promoting health applied is restricted to the severity of the conditions included, rather than the prevalence of these conditions. In this way, neonatal screening programmes seem to rest on a rationale that is notably different to the one apparently used for motivating screening in the prenatal area (see Section 3.1.1). As a consequence, neonatal screening programmes as they have traditionally been designed are considerably easier to reconcile with the traditional individualistic outlook of medical and health care ethics than prenatal screening.

At the same time, the fact that the tested individual is a newborn relates to an obvious downside of neonatal screening: those screened cannot autonomously consent to being tested themselves. As a general rule, it seems plausible to infer from the general idea of respecting patient autonomy the principle that, if one can expect a currently decision incompetent patient to become sufficiently autonomous in order to make a decision of her own regarding what health care measures to undergo and there is no significant harm to that patient by waiting for that to happen, then one should so wait. This principle is based on the idea of respecting autonomy by interpreting this respect to target the values and wishes of the patient. Thus, if there is no countervailing reason, when decision competence is not in place, but may be expected to be in the future, waiting for such competence in order to have

---

[31] This line of reasoning is of course in the spirit of Wilson and Jungner (1968), p. 27: even if prevalence is low, preventing very serious disease for a few individuals can justify screening. This, naturally, again gives rise to the question of how serious something has to be in order to be serious enough. We will return to this question shortly.

[32] Beauchamp and Childress (2001), pp. 102–103.

the patient decide herself is to honour the value of having this patient direct her life on the basis of her wishes and values rather than someone else's.

Understood in this way, the medical ethical rule of *respect* for the autonomy of patients will always provide *some* reason against neonatal screening. In practice, though, this reason will often be quite weak. The diseases targeted in standard neonatal screening programmes are very harmful, and even if the probability of a single baby to actually have any of these diseases is low, the risk of harm may still be judged as important enough to trigger the clause that a postponement of testing should not be significantly harmful. Moreover, some diseases tested for have serious impact on the future *autonomy* of the infant if not treated, for instance PKU and biotinidase deficiency. This is an argument for neonatal screening that assumes the *promotion* of future autonomy to be an important reason for such a practice (although seldom mentioned in the motivation of existing programmes). However, if early detection is not an advantage, respecting autonomy speaks in favour of postponing testing until the patient is old enough to take a stand herself. Since the primary beneficiary (or patient) in neonatal care is the newborn child, it is thus plausible to maintain the focus on advantages of *early testing* in terms of the well-being and – to some extent – the future autonomy of patients.

3.1.2.2 Neonatal Screening and Parental Informed Consent

The fact that the person tested in neonatal screening is not capable of giving informed consent also actualises the issue of to what extent *parents* (as is often the case in paediatric care – at least to a certain extent) should be given the role of proxys. That is, even if there are apparently good reasons in terms of health to screen newborns for certain conditions, should informed consent procedures be applied to the parents – thus giving them an opportunity to withdraw their child from the programme? In actual programmes, standard practice in most developed countries nowadays is to give them such an option, but historically more or less clearly mandatory participation has been practiced and is so still in some states of the USA.[33] At the same time,

---

[33] In particular, for apparently no good *health care ethical* reason, several states still recognise only religious objections as a valid reason for parents to withdraw their children from screening and a few states do not recognise *any* reasons as thus valid. Personal information and correspondence from Amy J Hoffman and the Bioethics and Legal Issues Workgroup of the Newborn Screening Translational Research Network, USA. Further, though not completely up to date, information can be found in The President's Council on Bioethics (2008).

as observed in Chapter 2, informed consent procedures can be more or less biased towards having people enrol in or withdraw from screening and, we conjecture, in the absolute majority of countries providing parents with an opportunity to withdraw their children from neonatal screening, the consent procedure is heavily rigged in the former manner. Is any of this acceptable?

This issue used to be relatively easy to tackle, since either lack of participation was not really an issue of practical concern, or the tests offered by neonatal screening programmes used to be of such obvious and unequivocal potential benefit to children that the criteria generally applicable to paediatric clinical practice for overruling parental objections would seem to be valid also for neonatal screening and similar well-motivated public health oriented interventions (Dawson, 2005). This reasoning is strengthened if we also allow the additional reason for neonatal screening in terms of promoting the future autonomy of the child (since the reduction of parental autonomy is then motivated by protection of the future autonomy of another person).

However, the rather drastic expansion of neonatal screening that has taken place in many countries since the 1990s (to be discussed in Section 3.1.2.3 below) has made this picture less clear and provoked debate on the subject of parental informed consent. It is generally recognised that the expanded programmes make participation less obviously beneficial to children – partly due to the large number of conditions included and the wide variation among these as to symptoms, progression, prognosis and treatability, partly since new types of psychosocial side-effects, such as those of medicalization[34] and stigmatisation[35] have been highlighted. This also seems to imply that the expanded programmes are less clearly motivated from a public health perspective, so the idea of parental informed consent being of less importance due to public health concerns[36] does not appear to be obviously warranted.[37] At the same time, due to the very rich and complex body of information that may be revealed by the expanded programmes, to apply fully the ideal of informed consent to parents may seem to be both ethically and practically

---

[34] Verweij (2000).
[35] See further Section 3.1.3.
[36] Nijsingh (2007).
[37] It may also be added that even if we were to assume the programme to be clearly motivated in terms of population health, applying parental informed consent would constitute a problem from a public health perspective only if it led to a significantly lower uptake. From the overarching view of population health, the occasional drop out of one or a few individuals does not make much of a difference – in particular when the probability of each individual to test positively is as exceptionally low as in the case of the classic neonatal screening programmes.

3.1 Diseases and Groups

dubious and, furthermore, severely threaten the cost-effectiveness of neonatal screening.[38] On the other hand, the programmes are (just as prenatal screening programmes) nowadays less of separate public health undertakings and more of integrated activities in routine clinical health care, thus making it problematic to deviate from informed consent standards otherwise applied within this institutional context.[39]

We will be returning to the issue of parental informed consent in Section 3.1.3 and then revisit the consent issue in both the neonatal and other screening contexts in all of the following chapters. At this stage, we merely want to highlight three aspects of immediate concern. First, we hold it to be a rather incomprehensible and remarkable position to allow for parents to opt out of neonatal screening on behalf of their children *for religious but not for other reasons*. Nevertheless, this is the current policy of, e.g., the state of Michigan in the USA, where the relevant section of the Public Health Code reads:

333.5113 Medical treatment, testing, or examination as violative of personal religious beliefs; compliance with provisions regarding sanitation and reporting of diseases.

Sec. 5113.

(1) Except as otherwise provided in part 52 and section 9123, this article and articles 6 and 9 or the rules promulgated under those articles shall not be construed to require the medical treatment, testing, or examination of an individual who objects on the grounds that the medical treatment, testing, or examination violates the personal religious beliefs of the individual or of the parent, guardian, or person in loco parentis of a minor.

(2) This section does not exempt an individual from compliance with applicable laws, rules, or regulations regarding sanitation and the reporting of diseases as provided by this code.[40]

As far as we can see, none of the health care or public health ethical reasons that may support parental informed consent in neonatal screening would restrict this support only to a certain class of reasons – be they religious or of some other kind. The type of statute exemplified above reminds of the old days when it was apparently (but falsely) thought among many policy makers that any serious moral objection that a person might have against participating in military service would have to be religiously grounded. We

---

[38] Nijsingh (2007). The problems, perils and demands of counselling are further addressed in Chapter 4.

[39] Friedman Ross (2011).

[40] State of Michigan (1978). We owe this particular information to Denise Chrysler (University of Michigan School of Public Health).

recognise, of course, that there are historical and cultural *explanations* for this type of irrational bias in favour of religious thinking. However, from an ethical point of view – regardless of what basic ethical position is assumed – these explanations do not amount to justification. On the contrary, to the extent that there are ethical reasons to apply informed consent, these reasons hold equally for all decision competent people, regardless of what reasons they may happen to use to decide whether or not to consent. To claim otherwise would be to question the very basis of the informed consent idea – that of there being of value to respect the autonomy of all people equally.[41]

Second, the idea that, while not applying informed consent procedures, parents may still be given the opportunity of withdrawing their children from neonatal screening,[42] is to our eyes the least attractive position available on the topic of whether or not parental informed consent should be applied or not. This since it is so blatantly in opposition to the notion of acting in the best interest of children: the idea essentially implies the right of parents to block their children's access to potentially beneficial procedures for no reason as regards these children's interests while denying the right of doing the same but on the basis of some reason. Such a policy achieves nothing in terms of protecting or promoting the autonomy of anyone. Indeed, as we will be arguing in Chapter 4, counselling is an important tool for screening programmes to meet not only autonomy-related requirements, but also the standards of health improvement aimed for. In the case of small children – whether or not parents *consent* – they need to be properly informed in order to be able to provide the assistance and support on the homefront as a rule required in the long-term care of seriously ill children.

Third, it is a rather striking feature of the debate about parental informed consent in neonatal screening that it tends to take the above mentioned expansion of such screening programmes as a given prerequisite. In contrast, later on, we will argue that difficulties to apply without serious side-effects sufficient counselling and informed consent procedures may many times provide powerful reasons against screening in the first place – especially so regarding programmes where the benefits for individuals are less obvious.[43]

---

[41] Cf. Nijsingh (2007).

[42] This idea seems to be at least implicitly implied by Nijsingh (2007) when he opposes the ideal of full informed consent in neonatal screening mainly on the basis of the problems of meeting the information requirement of this ideal. This seems to be compatible with the design of not applying that requirement while still leaving room for parents to have their children opt out of the programme.

[43] Nijsingh (2007) claims that such an idea violates the idea behind the ideal of full informed parental consent since "in this case the parents' choice is being pre-empted

3.1 Diseases and Groups 49

3.1.2.3 Expanding Neonatal Screening – How Far?

Even if the ethical basis and goals of neonatal screening are less controversial than regarding prenatal screening, and the established programmes seem well entrenched in both health care and public health, new developments have provoked quite some controversy. This is due both to the already mentioned development of new testing methods and the fact that the ethical basis of neonatal screening is not as clear-cut as may be suggested by first impression.

The method of tandem mass spectrometry, a technology which makes it possible to screen for a large number of inborn disorders, has attracted much attention.[44] This method not only allows for detection of considerably more such disorders than traditional clinical diagnosis, but also is rapid, relatively inexpensive and can be used to screen for a great number of disorders simultaneously. It is thus a good example of how medical and technological development makes it unavoidable to address the question of which disorders to screen for among those where it is technically and economically feasible.

Many groups and experts have taken on the task to answer this question. However, despite apparently similar points of departure, they have reached different conclusions. This, in turn, depends, at least partly, on the consistent use of rather imprecise terminology. For instance, the following criteria for which conditions that are legitimate candidates for neonatal screening are fairly representative:

- It can be identified at a phase (24–48 h after birth) at which it would not ordinarily be clinically detected;
- A test with appropriate sensitivity and specificity is available for it;
- There are demonstrated benefits of early detection, timely intervention and efficacious treatment of the condition being tested.[45]

Even though well in line with the traditionally accepted goals of neonatal screening and even though agreed upon on a general level, the

---

too" (p. 211). However, this argument seems to rest mainly on ignoring the distinction between the ideals of respecting and promoting autonomy. The informed consent ideal is based on the former, and respect for autonomy does certainly not require that health care offers all it could possibly offer to people (for them to consider in terms of choosing to have or not), but is quite consistent with limiting these offers to procedures that are judged to be well-motivated from the point of view of health.

[44] See e.g. Pandor et al. (2004); and Wilcken et al. (2003).
[45] ACMG (2005), p. 5.

vagueness of the terms used in these criteria makes much room for different interpretations.

Take for instance the expression "efficacious treatment". How should this term be understood? Even if one interprets it in terms of generally accepted standards of health, such as morbidity and mortality, there is still the question of how substantial the gains in terms of these variables must be in order for screening to be justified. Similar questions will remain in the case of expressions such as "appropriate sensitivity and specificity", "demonstrated benefits", and so on. Our main point here is that all these remaining questions are *ethical* rather than medical, and cannot be settled by any number of empirical investigations. To be sure, the ethical assessment would benefit very much from access to well-founded facts about the effects of the available treatment procedures, etc. However, a problem with regard to this is that many of the disorders that have been proposed for neonatal screening with the new methods are extremely rare.[46] Because of this, well-designed randomised controlled studies of the disorders and the effect of diagnosis and treatment have not been made and are often difficult to make at all.[47] This relates to another ethical question: how "demonstrated" must the benefits be, in terms of, e.g., treatment, in order to be judged as well-founded enough? Plausibly, this issue has to be addressed in connection to the question of how serious a disease must be in order to be a candidate for screening, and the question of how much prevalence should matter,[48] but *how*? This query opens up a complex of underlying issues in medical ethics with regard to the ethical assessment of risk that, we suggest, have not been sufficiently analysed and debated.

In addition, it has transpired that the application of tandem mass spectrometry not only leads to detection of the conditions actually targeted, but also to a large amount of so-called secondary findings. What is more, several of these do not meet some of the conditions above.[49] It is unclear, therefore, if these are viewed as covered by the above criteria (in which case the expansion of screening becomes much more difficult to justify), or if they are seen as irrelevant side-effects.

---

[46] For 12 of the 29 disorders proposed for a general nation-wide screening programme by ACMG, there is an incidence of less than 1 in 100,000 live births. Natowicz (2005), p. 868.

[47] Natowicz (2005); and Wilcken et al. (2003).

[48] Here, the ACMG played down the importance of prevalence to an unusual degree (Natowicz, 2005).

[49] Personal information from Ulrika von Döbeln.

3.1 Diseases and Groups

So, depending on how the criteria and goals are interpreted in detail, policy and practical recommendation for screening can vary widely. It should thus come as no surprise that while ACMG recommends 29 disorders for newborn screening (that become 49 if secondary findings are included),[50] a report for the Health Technology Assessment Programme of the National Health Service of the United Kingdom recommended only 5.[51] Most other European countries lie somewhere in between these recommendations and the gap between Europe and USA is closing[52] – albeit still remaining significant.[53] This reflects a difference in the *evaluation of* data rather than a difference of data and, we propose, underlying these different evaluations we find ethical issues of uncommon complexity that are as a rule not made explicit.

Another underlying explanatory factor of the differences between the UK and the USA instead connects to differences with regard to socio-economical and political structure: in the USA, neonatal screening has as a rule been organised on state rather than federal level, which has brought substantial differences with regard to the diseases screened for in different states. This, in turn, has effected lawsuits from parents and patient advocacy groups in states with more restricted programmes. Due to the way in which the legal system interacts with health services mainly through civil law, involving also the powerful private health insurance industry, these actions have put heavy pressure on professionals to expand the programmes (in order to escape astronomical compensation claims). Compare this with European countries, where there is usually a unified national health service that interacts with the legal system mainly through administrative and criminal regulation. The importance of mentioning this type of aspects is that they serve to highlight, once again, how many ethical issues connected to screening have to be considered in a wider societal

---

[50] ACMG (2005), p. 16.

[51] The report only recommends adding MCAD deficiency to the standard neonatal screening programme (Pandor et al., 2004, p. 113), which is already implemented.

[52] For instance, Sweden introduced screening for 19 new disorders, including secondary findings, in November 2010 (http://www.karolinska.se/Karolinska-Universitetslaboratoriet/Kliniker/CMMS/PKU-laboratoriet/Information-om-nyfoddhetsscreening/).

[53] See previous footnote. One telling difference between USA and Europe is screening for SCAD, which is included in the AMCG recommendations, but which was removed from UK screening programmes, since it presents such rare and often mild symptoms that it is unclear whether it should be considered a disease at all. Accordingly, e.g. Sweden and Denmark have not included SCAD among the conditions generally screened for (personal information from Ulrika von Döbeln).

and cultural context. In the case of prenatal screening, we pointed to the important role of widespread politico-cultural views on disabled people and how they may be treated in society. What we see in the case of neonatal screening is a parallel phenomenon, but here connecting rather to societal traditions with regard to legal systems and the organisation of health services. Below, we will encounter other aspects of the importance of a wide scope on the ethical discussion, when considering screening of adults and children.

Whatever future size neonatal screening programmes will take, on the basis of what has been said in this sub-section, there is reason to highlight two issues of importance, besides the one already mentioned regarding the ethical aspect of the assessment of underlying evidence regarding the health-related effects of screening. First, as will be addressed in more detail in the next sub-section, expanding neonatal screening programmes to include less severe conditions with more uncertain treatment possibilities makes it less improbable that parents would want to opt out of the programme. This since the benefits to the child of being tested become less obvious. At the same time, less saliently severe conditions and more unreliable treatments creates an uncertain situation for the parents as to whether to trust the information that their child is or is not healthy. Unlike before, neonatal screening will then have to grapple for real with the issues of how to value parental consent and how to assess the risk of stigmatisation, already briefly commented on in Section 3.1.2.2, this issue will be further discussed in Section 3.1.3.

Second, the inclusion of a wider set of conditions seems to mean that the neonatal screening practice starts to slide with regard to its ultimate rationale. As we have seen, traditional neonatal screening seems to be firmly placed within a traditional health care ethical context in terms of the best interest of the patient. In contrast, prenatal screening, with its emphasis on prevalence rather than severity of conditions, fits better within a public health ethical framework – something that, as was seen earlier in this chapter, has sparked sharp and, to some extent, well-founded criticism. When neonatal screening programmes include more conditions on the basis of prevalence rather than severity and treatment prognosis (as would be the case with screening for fragile X, see 5.2), it moves ideologically closer to prenatal screening, thus opening itself up to similar criticism. This does not imply that the expansion of neonatal screening discussed earlier thereby *is* justified on the basis of such an ideology – the many uncertainties as regard treatability, various side-effects and the costs of ethically required counselling makes this rather uncertain. However, the shift of ethical focus is still of importance to note.

3.1.3 Child and Adolescent Screening

Under this heading we will consider some special issues that arise from screening programmes targeting children after the neonatal period, including teenagers up to the age of 18 (or whatever age at which one is legally considered as an adult). In general, several of the considerations that have already been brought forward are as applicable to the screening of this target population. For this reason, we will mainly highlight issues of particular importance for the area of screening of children and adolescents. It should be underlined, however, that these issues regard not only programmes targeting children and adolescents exclusively, but also any screening where children or adolescents may be included in the target population.

As in the case of neonatal screening, the reason for screening children or youngsters can mostly be cast in terms of traditional health care goals – although much such screening may also appear as motivated from the perspective of public health. The latter regards, e.g., the sort of programmes for the routine checkup of children's physical and mental development that are applied in many countries, as well as similar programmes for dental health. It may be noted that such programmes belong to the classic toolbox of public health practice and that they as a rule involve treatment not only in the form of reparative measures when problems are discovered, but also purely preventive measures benefiting also those people where no problems are discovered, such as education about how to maintain one's good health.

At the same time, the problem with regard to lack of an autonomous decision maker also reappears. As a general rule, therefore, it seems that the requirements of screening programmes targeting children and adolescents should be similar to those of neonatal screening. This makes the ethics of this area of screening rather restrictive, although it also means that all of the more difficult ethical issues raised by neonatal screening are relevant in the case of children and adolescents as well. In particular, it is worth noting that while there are many cases where testing of individual children on the basis of some special reason for worry may be warranted,[54] this does not justify programmes aiming at having children *in general* tested for the conditions in question.

However, compared to neonatal screening, adolescent screening programmes also give rise to some especially complicated issues. One of these has to do with the apparent fact that, more so than in the neonatal case

---

[54] See, e.g., Robertson and Savulescu (2001).

(as noted above, at least regarding the traditional neonatal programmes), in adolescent screening it may be expected that some parents will be unwilling to have their children entering the programme. This raises the question of the rights of parents to deny their children potentially beneficial health care measures.

On the one hand, one may want to deny them such a right with reference to the best interest of the child and the fact that the case for respecting parental autonomy does not include respecting any wish parents may have with regard to their children.[55] On the other, one may point to the fact that there are many areas where parents *are* allowed to make decisions on behalf of their children that may be detrimental to the interests of the child. Just as in the case of childhood vaccination, it is to be expected that this issue will be judged differently in different socio-cultural contexts.[56] In general, however, the following line of reasoning seems to hold: if parents are accorded such a right in neonatal screening, consistency requires that this right is extended also to the case of older children and adolescents. If parents are not given such a right in the neonatal case, whether or not they should have it when their children are older would seem to depend on the extent to which the values promoted by the screening programme are as salient and undeniable as in the case of typical neonatal programmes.

As mentioned in Section 3.1.2.2, the views and practices on this matter are shifting around the world. In practice, though, also neonatal programmes that do give room for parental informed consent apply some sort of middle ground: in principle, parents may receive quite a lot of information and counselling and eventually choose to withdraw their children from the programme, but the pressure not to do this is rather heavy. "Opting in" is typically presented as the default position – usually the test is presented as a more or less automatic ingredient of routine neonatal care aiming to promote the interests of the child. This serves to obscure the fact that parents may, in fact, if they inquire further, find reasons to choose not to have their baby tested. Although, in Section 3.1.2, we pointed to problems with the justification of the new expanded neonatal screening programmes, the original rationale for either denying parental informed consent or to rig its practical application heavily in favour of participation would seem to be cast quite easily in terms of both the best interest of children and concern for population health. In the case of screening children and adolescents, therefore, the consistent view would perhaps be that the better reasons there are for

---

[55] Dawson (2005); and Nijsingh (2007).
[56] Dawson (2005); and Moran et al. (2006).

the programme in terms of such reasons, the more acceptable it is to accept pressures on parents to join, similar to those applied in the neonatal case.

Again, the factors highlighted earlier regarding the severity or prevalence of the disease, the quality of the test and available treatments will be crucial to consider. However, when, in the case of children and adolescents, these factors are not as clear-cut as they are in the classic neonatal programmes or the routine physical development checkup programmes mentioned above, further factors are brought to the fore. This may regard, e.g., suggestions for having the checkup programmes include equivocal psychological status categories, such as ADHD, where the success of available treatments are less obvious.[57] Another example may be the application of presymptomatic genetic testing for delayed onset diseases available for some groups of adults also to children.

3.1.3.1 Stigmatisation

When the diseases targeted are more complex and/or the tests less unequivocal or difficult to comprehend in relation to the child's actual health status, the information provided by the test threatens to affect the parents in a way that is harmful to the child. In particular, the phenomenon of so-called *stigmatisation* has been suggested to be such a risk.[58] What is meant by this is that a piece of information regarding the expected future health of the child makes the parents alter the way in which they relate psycho-socially to it. For example, they may start to perceive their child as much more unhealthy and vulnerable than what is actually the case (they, so to speak, cash in a future negative outcome beforehand). This, in turn, threatens to seriously upset the parent-child relationship in ways harmful to the child, not least by impeding the room for personal growth and maturation. Most seriously, this type of effect may occur even if the information only suggests a *risk* of future health problems – a risk that may in fact never be realized. Similar effects are to be expected when treatments are less certain regarding prognosis, since the parents will then feel less sure about the effects of treatment interventions, even if they are in fact successful. In relation to screening, the stigmatisation threat is extra salient, since the parents will, as a rule, not be very well

---

[57] It should be underscored that we are not here putting into question using the detection methods that might be employed for such programmes for the purpose of "screening" in the course of epidemiological research. On the contrary, such research may provide valuable data on the prevalence of such conditions, as well as the actual success of various treatment attempts (Rowland et al., 2001).
[58] See, e.g., Ablon (2002); Levy (1980); Parker (1983); and Thomasgard (1998).

prepared to receive a possible message about expected ill-health, uncertain prognoses, complicated treatment scenarios, and so on.

The stigmatisation may also regard how the information from the test tends to sipper into a wider social context. A stigma within the family can be expected to be communicated to acquaintances, relatives, day-care and school staff, et cetera, thereby affecting the child also in those settings. Furthermore, due to politico-economic surroundings, the stigma may acquire the characteristics of outright discrimination and affect, e.g., the child's future access to employment or health insurance. Such scenarios will be further explored in connection to adult screening in Section 3.1.4.

3.1.3.2 The Child as Decision Maker

Screening of children and adolescents is further complicated by the fact that children are in a permanent state of flux as to decision capacity and may, in different developmental stages and ages, be more or less mature and capable of understanding what a testing offer is about and form a reasoned opinion about this offer. They may therefore be more or less close to the ideal type of an autonomous decision maker, whose autonomy should be respected according to standard medical ethics. From a public health ethical perspective, as mentioned in Chapter 2, the importance of the rule of respecting autonomy is not as obvious, but there is room for asking the question when a public health based concern for autonomy on the population level[59] would become applicable to children.

This fact gives rise to a number of issues. One of these connects to the former point regarding parents: plausibly, the standard view of law to have a rigid cut-off point at a certain age, where parents lose all their rights to rule the lives of their children, cannot be justified in the context of a more careful ethical analysis (even if it, for reasons of practicality, must remain the standard tool from a legal point of view). Rather, from the point of view of respecting autonomy, the case for letting the wishes of the child influence the decision becomes stronger the more mature the child is and it may very well be the case that a child younger than the legal cut-off point is mature enough to be a fully autonomous decision maker, whose autonomy should thus be respected according to standard medical ethics. And, since the child and its parents need not want the same thing, the stronger such reasons there are, the more probable it is that balancing conflicting wants of the child and

---

[59] E.g., in the form of a goal of public health to promote the equal health promoting opportunities of all, as proposed in Munthe (2008).

3.1 Diseases and Groups

its parents becomes a serious ethical problem in the context of screening of children and adolescents.

Now, if to the idea of respecting autonomy is added the suggested goal of *promoting* autonomy, the just mentioned issue acquires a further dimension that was briefly touched on in Chapter 2. If a child of – say – 13 years of age wants something that threatens her future development into a fully autonomous individual, the value of promoting autonomy may counterbalance the reason to respect this want. Even if there is some hesitation to accept infringements of the liberty of adults for the reason that this will promote their future autonomy, it is quite common to justify coercion or manipulation of children (e.g. in upbringing) from the alleged fact that this will further their development into decision competent adults. In the case of adolescent screening, this type of consideration would seem to work in tandem with ordinary health reasons, since the possible autonomy-related benefits in this context can be expected to parasite on the benefits involved in avoiding, curing or ameliorating the effects of disease. However, the consideration still adds one further factor that, if accepted, may strengthen the case for overriding the wants also of more mature children and youngsters in situations where a screening programme brings large and clearly demonstrated health benefits.

At the same time, the risks of stigmatisation and discrimination mentioned earlier affect not only the well-being of the child, but also its chances both to grow and mature as a person (which may be impeded by an overprotective social environment) and, when older, encounter obstacles when trying to execute its life-plans in a socio-economic context that brings risks of discrimination regarding employment and insurance. In virtue of such risks, therefore, there may be reasons to doubt that screening will in fact promote the child's future autonomy, even if considerations of health alone point in that direction.

Of course, all of the considerations mentioned have to be balanced against other reasons in favor of a screening programme. Some programmes may very well be justifiable even in the presence of some risks – for instance, if they contribute substantially to the promotion of population health. What is more, however, the risk of stigmatisation, as well as the reason to let the children themselves have some say regarding the participation in screening programmes, both point to the importance of how the programme is organised and followed up. In particular, it highlights the crucial importance of counselling as a way to minimise these types of risks and meet these kinds of requirements. We will return to that particular issue in Chapter 4. Another aspect, mentioned in connection to prenatal screening, is the nature of the wider socio-cultural and politico-economic context in which the screening takes place. In the prenatal case, what was highlighted was how prenatal screening (in contrast to testing not organised as screening) seems to feed into and reinforce existing devaluating attitudes and discriminatory practices towards sick and disabled people. In the present one, something similar is brought to the fore when it comes to the issue of whether or not the benefits of screening programme that targets children and/or adolescents is strong enough to warrant the risks involved. We will return to this aspect in Chapter 6.

3.1.4 Adult Screening

When considering adults, several of the *theoretical* problems with regard to respecting and/or promoting autonomy highlighted in earlier sections seem to disappear. At the same time, though, since adults are considered to be the paradigm case of a competent decision makers, we instead face acute *practical* problems regarding how to organise screening programmes in order to actually respect and (if that is considered desirable) promote autonomy. This raises issues about the quality of the information that can be provided by the test used in the programme, as well as how much and what types of counselling procedures are needed. We will return to these issues in Section 3.2 and Chapter 4 below. For now, it suffices to underscore that, in many cases these aspects cannot be perfectly isolated from the similar ones arising in other contexts, such as the prenatal one (since the test results of an adult may have bearing on the expected health of this person's offspring).

Besides this, screening of adults mainly raises the basic ethical issue of whether or not such a programme would be "good enough" in terms of its potential benefits and risks. Mammography screening programmes for breast cancer and screening for prostate cancer using the so-called PSA method are two debated examples in this respect.[60] As before, the issue of the balance of potential benefits and risks may often be analysed in terms of ordinary health care or public health goals, thus leading back to the value of health. However, sometimes it may be more adequate to analyse the benefits (and risks) of screening programmes targeting adults mainly in terms of autonomy. For example, as mentioned, prenatal screening may be described as targeting potential parents (rather than foetuses), and seen from that perspective it has very little to do with future health. Assessments of these issues will furthermore give rise to the complex task, highlighted above, of balancing chances of benefits, risks of harm and the varying quality of the evidence

---

[60] Both of these examples will be discussed in further in Sections 3.2, 5.3, and 5.4.

3.1 Diseases and Groups

underlying the estimates of these risks and chances. Also the adult case, therefore, points to the need for penetrating further the basic ethical query of what would be a proper performance of this task.

In addition to these familiar issues, the case of adult screening gives rise to practical considerations that underlines the need for ethical assessments of screening programmes to consider a wider socio-political context. These considerations have their roots in the fact that the information provided by the tests used in screening programmes may be of interest not only for health care and the individual, but also for various third parties. In the debate on the ethics of genetic testing, the most obvious such party identified is the blood relatives of a tested person.[61] Here, however, we will mainly focus on more distant and, at the same time, powerful such parties – in particular insurance companies and employers.[62] Also regarding those parties, the debate on the ethics of genetic testing provides a fruitful starting-point, although, as we will see, the general issue concerns any sort of medical information revealed by a screening programme.[63]

Genetic information is of interest to insurance companies and employers in that it may provide a useful tool for economic risk assessments regarding potential insurance holders and employees respectively. This utility is most salient in the case of insurance companies, in which case not having access to the information provided by a test, may make the company vulnerable to, what is called, adverse selection. That is, an insurance company's policy holders may have a significantly higher risk of contracting whatever health problem covered by the insurance, than what is reflected by the premium paid (since these holders know things from the tests not known by the company). Similarly, an employer may have to pay as much salary, although some of the employees have weaker health than the staff of the competitors. So, both insurance companies and employers have a general interest of avoiding costly business relationships, and both may increase their chances

---

[61] Juth (2005), chapter 6.

[62] As noted in passing in earlier sections, the interest of insurance companies is of some relevance also to the testing of foetuses and children. One very popular insurance product is health insurance for children, and most such insurances are underwritten only if the health risks to the child is assessed to be low enough (alternatively, the premium is bearable for the parents only in that case, or compensation is paid only if a health problem does not have its origin in physiological features revealed by earlier tests or are simply inborn).

[63] For a more comprehensive treatment of this issue with regard to insurance companies, see Radetzki et al. (2003). The same with regard to employers is found in Juth (2005), section 7.6.1.

of such avoidance by having access to health information about potential customers or employees.

This interest is not confined to *genetic* health information, but really concerns any information that may provide basis for predictions of future health. Thus, the problem highlighted by the debate about genetic testing extends to most sorts of medical tests. For instance, this aspect has been a theme in the debates connected to the recent US reformation of its Medicare programme.[64]

The importance of this for the ethics of screening is, of course, that the just related facts complicate the assessment of whether or not a screening programme is "good enough" in view of its goals. Even if the goal is only to promote health, failure to realise this goal may result from the fact that a person is in the future denied job or health insurance due to the information revealed through the programme. And, in view of that, respect for autonomy seems, at least, to require that the information and counselling in connection to screening programmes make such risks salient.

At a more overarching level, this complication widens the scope of socio-political issues of relevance to screening even further. For one way to avoid or minimise the risks just pointed out is, of course, to organise society in a way that accomplishes this (such as publicly financed health insurance good enough to make private health insurance unnecessary for having a decent quality of life).[65] To the extent that the practice of screening aims at promoting public health, such measures seem to reside well inside the area of interest for a plausible ethics of screening.

3.2 Investigation, Testing and Analysis

When considering the implementation of a screening programme, besides the properties of the disease and target population, one also has to ponder the properties of the methods for investigation, testing and analysis applied.[66] One such property that is particularly salient is the monetary cost. This is important since, as noted at the outset, the practice of screening is located in a public health context, where cost-effectiveness is a crucial consideration. More expensive methods will, thus, always add to the reasons against

---

[64] Brandon (2009).
[65] Juth (2005), chapter 7; and Radetzki et al. (2003), chapter 7.
[66] Wilson and Jungner (1968), pp. 30–31, thus discussed there being a "[s]uitable test or examination" as a principle for screening. See also Shickle (1999), pp. 30–31; and ACMG (2005), p. 5.

3.2 Investigation, Testing and Analysis

screening. Such reasons may, of course, be counterbalanced by various benefits of the programme. However, the possibility of such benefits will, in turn, depend on other properties of the testing method. Below, we will address three aspects with regard to this: safety, validity and predictive value.

3.2.1 Safety

The way in which the information provided by a medical test is collected may be more or less risky. The examples of routine developmental checkups of children usually employ a mixed battery of methods, involving manual physical examination, blood and urine tests and (in the case of dental care) x-ray imagery. In genetic and many other types of testing of adolescents and adults, the method often consist of an ordinary blood sample, or even completely non-invasive ways of collecting tissue. As long as standard sanitary requirements are observed, the risks associated with these methods can plausibly be considered negligible.[67] A contested case is mammography screening for breast cancer, where the frequent use of x-ray may be held out as risky. In this and several other cases, other aspects of the tests may also be suggested to constitute downsides in terms of safety due to elevated risks of overtreatment. Such aspects will be further discussed in Section 3.2.2.

In contrast to the typical testing methods used in the screening of born people, prenatal testing often involves amniocentesis or chorionic villus sampling, in which biological material from the foetus is extracted from amniotic fluid (collected invasively from the uterus) or the placenta (likewise invasively biopsied). In particular, such invasive methods are required to ensure a high degree of precision and reliability of the test result. These procedures both bring a 0.5–1% risk of miscarriage.[68] Even if *pure* screening programmes involving any of these methods in the first place seldom have been proposed, programmes addressing a sub-population of pregnant women over a certain age have been rather common.[69] Moreover, screening

---

[67] As regards the use of x-ray in routine developmental dental checkup programmes, a further condition is that the checkups are not done too often.

[68] Connor and Ferguson-Smith (1997), pp. 201–202.

[69] In some rare cases, like Sweden, the regulation of this practice say merely that these women should be informed about their age-related risk and the possibility of requesting prenatal diagnosis (*The Genetic Integrity Act*, chapter 4, section 1). In many countries, however, the approach towards the woman is supposed to be much more suggestive than this. And even in the Swedish case, there is anecdotal evidence that actual practice is more directive than the guidelines suggest.

programmes that would involve invasive methods as a follow-up procedure in order to ensure diagnosis are often considered in reproductive care. For instance, carrier detection for cystic fibrosis has been suggested, as well as for fragile X. Moreover, aberrations observed in routine obstetric ultrasound often lead to an invasive prenatal test. Of course, when securing diagnosis of the foetus involves a test associated with the mentioned risk, the pressure to test oneself involved in any screening programme (see Section 2.3.2) becomes particularly problematic from an ethical point of view.

This situation has led to the development of alternative methods for more specific and precise risk assessment not associated with any risk of miscarriage, for example blood sampling of the pregnant woman and subsequent analysis of a combination of serum indicators. Taken together with other indicators, such as special forms of ultrasound and the age of the mother, this method ensures that the adequacy of identified "high risk" pregnancies for Down syndrome increase up to about 90%.[70] In consequence, many pregnant women who would have been identified as "high risk" on the basis of age alone, and in many cases would have exposed themselves to the risk of invasive prenatal testing as a result, will be classified as "low risk" with this new method. Also, some women below the commonly applied "high risk" age threshold will be identified as "high risk", thus gaining reasons for considering an invasive test that would not have been provided by age alone. The recent introduction of this new approach in many developed countries has been reported to effect a significant decrease in the number of requests for invasive prenatal testing.[71] At the same time, however, the information provided by the new methods makes the totality of the information related to the pregnant woman much more complex and difficult to handle. We will return to the complications associated with that in Chapter 4. As indicated earlier (Section 3.1.1), however, there seem to be quite strong ethical reasons against organising the offer of these new risk assessment methods as *screening*, since the aim of reducing unnecessary invasive prenatal tests may be reached with a less ethically risky way of organising the offer (namely, to offer risk assessment to all pregnant women who request invasive prenatal testing, rather than to all pregnant women). We will return to issues connected to non-invasive prenatal screening suggestions in Chapter 5.

In other cases, proposed screening programmes instead face the problem of ignorance or severe uncertainty regarding the risks of the testing method, as is often the case when the method utilises novel technology.

---

[70] Saltvedt (2005).
[71] Nadel and Likhite (2009).

A good example is provided by the procedure called PGS or, sometimes, PGD-AS – a method utilising techniques of preimplantation genetic diagnosis in the context of overcoming infertility problems in assisted reproduction, and proposed as a routine part of in vitro fertilisation (IVF).[72] Briefly, this method involves invasive sampling of cells from IVF embryos at the germ-line stage. Although no short-term detrimental effects on these embryos have as yet been observed, since the procedure is applied at the germ-line stage, subsequent long-term side-effects, even in the second generation or later, cannot be ruled out. At the same time, in spite of initially strong theoretical reasons for believing so, it has become unclear to what extent this method really does its job of improving the efficiency of IVF; different studies indicate contrary results, with recent, larger controlled studies putting the idea into question.[73] This type of situation gives rise to the issue of how to take into account the quality of available evidence (or, rather, lack thereof) underlying the assessment of risks and benefits associated with the testing method when determining whether or not to implement a screening programme.[74]

3.2.2 Validity

The most discussed property of tests and analytical methods in the screening context is otherwise their ability to identify correctly those, and only those, who have or will have a disease – what Wilson and Jungner called *validity*.[75] A method for investigation, testing and/or analysis can fail to identify individuals with disease in two ways: by being positive (i.e. indicating disease or high risk thereof) when there actually is or will be no disease (*false positive*) or by being negative (i.e. indicating absence of disease or low risk thereof) when there actually is or will be a disease (*false negative*).[76]

---

[72] Gianroli et al. (1999).

[73] For an overview of the situation, see Hernández (2009).

[74] See further: Munthe (1999), chapter 6.

[75] Sometimes the validity of a test has been referred to as its *predictive value* (see e.g. Wilson and Jungner, 1968, p. 21). However, as will be seen below, predictive value is a more inclusive feature that is affected by the validity, but also by other factors.

[76] In addition to validity, as test's *reliability* – i.e. its level of accuracy in measuring whatever it is that it measures (often the presence of some bodily substance or biochemical process) – is, of course, also of importance. However, in the following, we assume that applied methods are adequate in this basic sense – where they not, using the method for screening would be highly questionable to start with.

Evidently, the validity of the testing and analytical methods used in a screening programme matters much for its quality. However, *how* the quality is affected is a more complicated matter. In the next subsection, we will consider particularly complex cases where the goal of a screening programme is to promote autonomy. Regarding other goals of screening, however, things are also complicated. This has to do with the fact that different effects are to be expected depending on whether the method applied is more or less likely to produce false negatives or false positives.

The likelihood of a test to produce a positive response in the case of a person who actually is afflicted by the disease tested for is termed *sensitivity*, and the likelihood of a test to produce a negative response in the case of a person who actually is not so afflicted is termed *specificity*. Thus, the sensitivity of a test determines the likelihood of false negatives (the more sensitive a test is, the less likely it is that a tested person receives a false negative result) and the specificity of a test determines the likelihood of false positives (the more specific a test is, the less likely it is that a tested person receives a false positive result).

Most screening programmes utilise methods that detect more or less indirect indicators of the actual disease. For instance, cholesterol is an indicator for increased risk of coronary heart disease: the higher level of cholesterol, the higher risk. Another example is provided from prenatal testing, where the presence of certain serum markers in the blood of pregnant women indicate probability of carrying a child with Down syndrome.[77] Whenever such tests are used, a threshold level has to be set in order to separate findings that should be counted as positive from those that should be counted as negative. So, for instance, the lower the level of cholesterol that is taken to indicate risk of coronary heart disease, the more test results will count as positive, and, for that reason, a larger proportion of the indicated positives will be false ones. In other cases (such as that of AFP and Down syndrome), the relation between the indicator level and the risk level is reversed, but that is of no consequence in this context, since a threshold that discriminates positives from negatives has to be set in either case. Similarly, when the threshold for a negative result is generously set, more false negatives will result. This gives rise to a general problem: the more the threshold is set to exclude all negative cases, the higher the risk of false negatives – that is, the worse sensitivity. And, similarly, the more the threshold is set to include all positive cases, the higher the risk of false positives – that is, the worse

---

[77] This is determined by the levels of AFP (amniotic fluid α-fetoprotein), chorionic gonatrophin (hCG) and unconjugated oestriol (uE3) in the mother's blood.

3.2 Investigation, Testing and Analysis

specificity. Thus, regarding these kinds of tests, there always must be a trade-off between sensitivity and specificity – one cannot increase the one without decreasing the other.

The question then arises what one should strive for: higher sensitivity or higher specificity? Both false positives and false negatives are associated with various downsides. False negatives have the apparent cost of creating a false sense of security. This, in turn, may impede or delay needed treatment, in the worst case with fatal results. On a societal level, this translates into considerable costs in terms of public health. However, false positives have their downsides too: unnecessary anxiety and stigma, as well as the burden of further investigations and/or the risks and hardships brought by unnecessary treatment. In the worst case, a person may suffer depression, social exclusion and physical harm by risky (but unnecessary) medical procedures. On a societal level, the higher the number of false positives, the higher the cost of unnecessary further investigations and/or treatment. Besides that, the trust in and legitimacy of the screening programme, and in the long run, of health care and public health institutions in general, may be damaged by many "false" answers, positive or negative.

The question of where to draw the line is thus difficult to answer without a closer examination of various other variables, e.g. the disease targeted, the consequences for the individuals tested as well as for society and health care in general, and the access of treatments. For instance, if a positive test result is associated with much anxiety, as is the case with many late onset genetic disorders, such as Huntington's disease and many hereditary forms of cancer, high specificity is desirable. Also, if diseases are connected with societal stigma, as sickle cell anaemia[78] or sexually transmitted diseases, avoiding the "disease-label" when there actually is no disease becomes increasingly important for the individual. Moreover, the case for the importance of high specificity is strengthened if there are no effective treatments, as in the case of Huntington's, or if existing treatments are very burdensome, as is the case with prophylactic mastectomy for preventing breast cancer, or abortion after prenatal diagnosis.[79] Prenatal diagnosis also provides an example of the fact that further factors may be relevant, such as the prevalence of the condition and the goal of the screening programme. Since the 1980s, prenatal screening programmes analysing AFP in blood samples from pregnant women with the aim of assessing the risk of foetal neural tube defects have been running in many countries around the world. In Sweden, however,

---

[78] Clarke (1994), p. 8.
[79] Sandén and Bjurulf (1988).

such programmes were eventually shut down, mainly due to false positives in an environment with a low prevalence in combination with support for the goal of having prenatal diagnosis promote autonomy and psychological well-being. In contrast, it seems more reasonable to opt for high sensitivity if there is an effective treatment and there are severe consequences for the individual in terms of health without early diagnosis, as is the case with neonatal screening for PKU and the other diseases targeted in the original versions of such programmes.

Traditionally, among screening specialists, high sensitivity has been seen as a more important feature of a screening programme than high specificity.[80] This is partly due to historical factors – early screening programmes were to a large extent concerned either with generally securing a decent level of population health (such as in the traditional developmental screening programmes for children), or with communicable diseases, such as tuberculosis, the effects of which could be devastating to public health if undetected. However, the practice of screening in developed countries has evolved into being increasingly about chronic diseases that result from life-style, sometimes in combination with genetic factors, such as cancer, diabetes and cardiac disease.[81] Many recent debates about screening have revolved around genetic disease and the genetic components of health problems. However, as we have seen, many genetic disorders cannot be effectively treated and are related to societal problems of stigmatisation and lack of goods such as insurance. Against this background, it should come as no surprise that the focus on avoiding false positives has increased.

Besides the socio-economic factors pointed out earlier, it is in this light that one has to understand the current controversies about proposed or ongoing screening programmes for different sorts of cancer, such as prostate or breast cancer. The quality of the methods for testing and analysis employed in such programmes – the so-called PSA test in case of the prostate and mammography in the case of breast cancer – have salient downsides in terms of both specificity and sensitivity. Thus, for instance, the PSA test leaves a significant portion of false negatives, thus threatening to lure people receiving a negative test result into feeling overconfident and fail to undergo prudent medical examinations and thereby risk missing the early detection of an actual future prostate cancer. Both PSA and mammography have been pointed out to produce a significant amount of false positives, associated with not only unnecessary anxiety, but also overtreatment using risky and

---

[80] Wilson and Jungner (1968), p. 31.
[81] Wilson and Jungner (1968), pp. 9–10.

burdening methods. For these reasons, screening for prostate cancer continues to be a highly contested idea unless the test and follow-up procedures are amended to considerably. Such amendments (complementing PSA with a rectal examination and ultrasound investigation, and significantly improved post-testing counselling procedures), however, involves an increased burden of testing on the patient and puts the public health rationale for a screening program into doubt due to elevated costs.[82] In the case of mammography, besides being burdened by the anxiety and overtreatment risks just pointed out, the actual testing method (x-ray) also imposes some risks by itself (of inducing cancer, for instance) when used frequently.[83] We will return to both of these examples and the controversies surrounding them in much more detail in Chapter 5.

Plausibly, the ethical assessment of how sensitivity and specificity should be balanced cannot be decided by a simple rule. Rather, such assessments would have to be done by careful consideration of each suggested screening programme in a broader context, where also other factors of relevance to the defensibility of a screening programme are taken into account. For this reason, we will now consider one such further factor in particular that relates to the test itself and is closely connected to its validity.

3.2.3 Predictive Value

The predictive value of a method for testing and analysis is its ability to deliver results that accurately, informatively and unambiguously predict the actual (future) health status of the tested individual. This feature of methods for testing and analysis is of high importance in a screening context, since the general idea in that case is to acquire results that can guide practical decision making (about further tests, treatments, life-style changes, etc.). For this reason, predictive value is crucial not only from the goal of promoting health, but also from the goal of promoting autonomy. This since worse ability of the information provided to guide decisions makes it less likely that a screening programme applying this method actually succeeds in helping people to make choices in accordance with their own plans and values. The predictive value of a test is, of course, highly affected by its validity. However, it is also affected by other factors that will be in focus in

---

[82] For a recent overview and further references, see, e.g., Denham et al. (2010).
[83] Gøtzsche and Nielsen (2009).

this subsection – factors that have the main effect of making the information provided *unclear* in its practical implications.

Problems regarding predictive value become accentuated when the results themselves are either constituted by population-based statistical figures rather than binary results stating the presence or non-presence of a disease, or based on such figures. This problem is a pressing one at present and in the immediate future, due to the increasing possibility of quantifying the predisposition for developing various diseases that are the result of both complex genetic and environmental factors, so-called *multifactorial* diseases. Many of the most common diseases belong to this group, such as Alzheimer's disease, diabetes, cardio-vascular disease, various forms of cancer, and schizophrenia.[84] Since other factors than one particular gene determines whether or not an individual will have the disease, a genetic test can only deliver a probability for having the disease. And, since this risk figure will in most cases be estimated on the basis of evidence that is probabilistic as well, the figure itself will often run a risk of being inadequate. Let us consider these two sources of uncertainty in turn.

First, one effect of the just said is that the risk figure provided by the test may vary between a very small risk for some disease to quite substantial risk levels. When the risk is small (as will be the case for single genes in most multi-factorial diseases), it may be doubted whether *any* information delivered by a test (positive or negative) can actually guide *any* decision making. Of course, also information regarding an *increase* in risk, like 30% increased risk of having diabetes before the age of 50, can lead individuals to reduce other risk factors, such as the intake of sugar. However, there may be difficulties in seeing the relevance of such statistical figures for oneself (especially so, if the risk level one is compared to is very low to begin with),[85] and even if one does, people often "oversimplify" statistic figures, leading to either under- or overestimation of risks.[86] And while underestimating risks can lead to failure of taking important steps in order to reduce risks, overestimating risks can lead to unnecessary anxiety, and the application of unnecessary and risky medical measures. Moreover, there is no guarantee that risk information leads to the "proper" reaction of reducing other risk factors; it could also lead to a fatalistic attitude: "Oh, it does not matter what I do, since I'm genetically determined to get ill anyway."[87]

---

[84] Connor and Ferguson-Smith (1997), pp. 157–161.
[85] Adelswärd and Sachs (2002).
[86] Shiloh (1996), p. 8.
[87] Shickle (1999).

In the light of these complications, a particularly problematic, but surprisingly common, practice is that not only medical scientists but also health service providers often present the information that can be provided by tests – in screening programmes as well as in the ordinary health care context – in terms of *relative* rather than absolute risk (i.e. a percentage that indicates an increase in risk level compared to some alternative state of things, rather than the actual risk level given the actual state of things). While this concept of relative risk is a highly informative and useful one in the context of medical *research*, in the *clinical practice* of testing and screening it is of doubtful value, especially from the point of view of autonomy. First, information about relative risk (such as the information that the presence of a certain genetic mutation increases the risk of developing melanoma before the age of 50 by 50%) is completely incapable of guiding decision making unless the absolute risk of people not carrying this mutation is given in precise terms. Second, even if that absolute risk is given (in the case of the p16 gene, for which testing is on offer, less than 1%), the formula of giving that risk figure plus the relative risk figure introduces unnecessary complications into the information, thus elevating risks of misunderstanding. After all, the point of giving these two figures is to facilitate the calculation of the absolute risk actually run by the patient, so giving this latter risk figure would seem to be a more straightforward an unequivocal way of communicating that information. Third, it is highly plausible to assume that most people are not aware of the nature of the difference between the concepts of relative and absolute risk, thus being prone to confuse the two. This last danger seems to us to be the most serious one, since it implies that people are likely to infer their risks to be much higher than is actually the case (in the case of melanoma just mentioned, people are thus likely to infer their risk to be 50% rather than the correct figure of 1.5% or less).

These are very good reasons indeed *not* to use relative risk in the presentation of either what sort of information that may be provided by a screening test, or when informing about the actual result of a test. What the patient needs to know both for deciding whether or not to enter a screening programme and what to do on the basis of the test result are the absolute risk figures. In spite of this, there is a widespread tendency in the practice of screening, as well as medical testing in general, for health professionals to press more strongly on relative risk figures when the absolute risks are low or uncertain. Most probably, this is due to the fact that holding out the relative risk means putting the focus on an impressive percentage figure (such as a 100% increase in risk), while focusing on the risk actually run would makes significantly less of an impression (e.g. for a 0.005% initial risk, a 100% increase in risk still is no more than a 0.025% risk in absolute terms). This tendency may be partly due to the tight links that often exist between clinical practice and medical research, making the thinking in terms of relative risk that is of value in the scientific evaluation of hypotheses "leak" into the clinical evaluation of patients and treatments. However, it would be naïve to disregard as a further contributing explanation the socio-economic pull of screening ventures mentioned at the outset of this book. Health service providers running screening programmes need to secure a sufficient volume of recruitment in order for the programme to remain economically viable or to present the screening effort as a significant endeavour for funding parties to become sufficiently impressed. Pressing on relative risk figures rather than absolute ones is a likely effective lever to those effects.

However, things are even more complicated than this. As a rule, several genes at different loci in the DNA determine the predisposition for multifactorial disease. This is well illustrated by a common type of diabetes, which is multifactorial.[88] Today 12 different genes are considered to affect whether a person will develop the disease or not. This implies 531,144 possible combinations of alleles. Since different genes have different penetrance and expressivity,[89] and since the environment also influences the risk of diabetes, a genetic test cannot even result in a precise risk-figure. Rather, if a useful test can be construed at all, the result it will present will be in terms of a range of risk, like 5–35% increased risk in comparison to the general population. If the risk-level is low to begin with, the predictive value of such a test, and thus the value as a tool for guidance for health care measures or private decision-making, is virtually non-existing. If the initial risk level is higher, such harsh a judgement may not be warranted, but it is still doubtful, to say the least, what use could be made of such a piece of information.

It must be conceded that the process of scientific and technical development is rapid in this area, and that already there are technical tools for simultaneous and rapid analysis of many different genetic loci. However, detection of no matter how many DNA-sequences makes for no useful medical information, unless there is a solid underlying body of knowledge that can use the result of this detection to produce individual health predictions of sufficient certainty and precision. The problem is that, due to the very fact of the astronomic complexity of the interplay between different pieces of the

---

[88] Juth (2005), p. 26.

[89] The penetrance of a gene is the extent to which the gene manifests itself in a population, thus determining how likely one is to have a disease given a certain mutation. The expressivity of a gene is the strength with which a gene manifests itself in an individual, thus determining which and how serious symptoms one will have given a certain mutation.

3.2 Investigation, Testing and Analysis

DNA and the environment indicated above, there will be a *very* long time before such a body of knowledge is available. In the meantime, the uncertainty of the information provided by genetic tests for multifactorial diseases will be even worse than indicated above, since whatever prediction is made, it will run a substantial risk of being wrong (due to the lack of sufficient evidence).

Of course, some diseases are more directly the result of a mutation within one single gene (so called *monogenetic* disease). However, even if tests for such diseases often give a more definite result, and thus have a higher predictive value, they are often still problematic in terms of predictive value from a screening point of view. This is especially obvious in cases where many mutations on the same gene can give rise to the same disease. For instance, over 500 different mutations can give rise to cystic fibrosis. To be sure, some mutations are more common than others in certain populations.[90] However, the risk of false negatives increases whenever some mutations are left out.[91]

Even if there are technical tools for checking many mutations at once, some monogenetic disorders require a previously positively identified relative in order for it to be possible to find the relevant mutation on the gene. One such example is BCRA1 gene, responsible for almost half of the known cases of hereditary breast cancer. For this disorder, a screening programme directed towards the general population is thus not feasible at all. Cascade testing, working oneself "outwards" from one identified case, is nevertheless possible. However, even though the disease is monogenetic, the gene has reduced penetrance, which means that only a certain percentage of the carriers of the gene actually will develop breast cancer eventually.[92] So, the risk for false positives with needless anxiety and unnecessary removal of breast tissue remains.

In other cases, the uncertainty of the result instead is about the degree of expression of clinical symptoms, age of onset, rate of progression, etcetera. For instance, in the case of Fragile X, the presence of the genetic mutation explaining this condition, there is a wide variation as to how the patient

---

[90] For instance, the mutation ΔF508 is by far the most common mutation among people in Denmark; the A455E is more frequent among Ashkenazi Jews (Gregg and Simpson, 2002, p. 333).

[91] Actually, the UK (and most american states) has neonatal screening for cystic fibrosis, a country which is in other regards rather cautious in its use of neonatal screening (see Section 3.1.2). There are some good arguments in favour of neonatal screening for cystic fibrosis (Wilcken, 2009). However, in light of the mentioned problems, the question is how it should be performed.

[92] The exact figure is contested (see Shattuck-Eidens et al., 1997).

is actually affected, stretching from some problems with impulse control, hyperactivity and social interaction to severe cognitive and social disability. In multifactorial diseases such as diabetes, the situation is similar with regard to individual variation, in spite of a common underlying biomedical problem. The phenomenon is present also with regard to chromosomal conditions, where the chief target of prenatal screening programmes, Down syndrome, is virtually legendary for its huge span of variability regarding physiological as well as psychological symptoms. It is thus safe to say, that the threat to the quality of screening programmes created by the fact that the methods for testing and analysis used provide unclear information potentially affects all types of screening programmes in terms of the groups and diseases targeted.

3.3 Treatments

As mentioned in Chapter 2, the most common ethical issue in screening with regard to treatments concern their efficacy and riskiness in terms of health and well-being. Available treatments may vary in this respect from well-established procedures that are highly effective with almost no risk of serious side-effects (such as the dietary treatment for PKU), to complicated and rather uncertain measures that bring a range of risks. As indicated earlier, it seems plausible to hold as a general view, that this factor has to be balanced against a host of other morally relevant features when assessing whether or not a particular screening programme would be defensible. For example, it has to be considered whether or not the properties of the test used bring risks of "overtreatment" due to false positives. However, in this section, we will merely comment on two further aspects of particular importance.

It is crucial not to underestimate the difference between ordinary medical testing and testing done within a screening programme in relation to the treatment aspect. In the first case, a particular individual with an initial problem or worry has approached health care to have the facts straightened out and, hopefully, the problem solved. The starting point is thus uncertainty with regard to what sort of information may or may not be forthcoming and, as a result, a substantial level of ignorance with regard to what sets of treatment options may eventually be relevant. The normal course of events is that decisions of what possible treatments to consider are made under way, when the initial uncertainty has been cleared. And, when doing so, the pondering of what treatments to offer can be made against the background of the particular situation of the individual patient, thus making possible various "tailor-made" solutions. In contrast, in screening, the programme

3.3 Treatments

is designed to search for a particular disease or risk thereof that has been defined beforehand among a large group of people that, as a rule, have no previous suspicion or worry. This suggests, first, that the general responsibility of health care to be able to offer an acceptable treatment at the outset is stronger in the screening case than in the ordinary health care situation. Second, when considering what treatment would be acceptable enough for screening to be justified, the possibility of "tailor-made" treatments must be discounted (since, to begin with, there are no identifiable individuals in relation to which a judgement of the possibility for such tailor-making can be made). Of course, one may construct the programme so that there is no specified treatment, but a more vaguely defined follow-up programme, where tailor-made treatment solutions are an option. This, however, is problematic in three ways: First, it is unclear if such a setup can live up to the treatment condition of Wilson and Jungner (rather, this solution seems to mean merely that there *may be* a treatment). Second, having a complex follow-up setup such as this is expensive and makes it less clear that the programme is able to motivate its costs. Third, solutions of this sort are problematic from an autonomy perspective, since, for the patient, it becomes rather unclear what entering the programme might mean.

The second aspect to be highlighted connects to the issue of what goals or basic values may justify a screening programme. As described in Chapters 1 and 2, not only is the issue of whether or not available treatments are good enough for screening to be warranted open for questioning, it may also be discussed what proposed follow-up procedures that may qualify as treatments *at all*. Pondering that latter issue, considering the basic values of screening becomes paramount. To illustrate the sort of issues that are brought to the fore by such an exercise, we will in the following elaborate on two cases of suggested treatments in the screening context that awake particular controversy – abortion and (mere) counselling.

3.3.1 Abortion as a Treatment

As described in Section 3.1.1, the idea of abortion as a treatment is found within the practice of prenatal screening. As pointed out in that context, this idea is highly questionable if the sole goal of the screening is to improve the health of the foetus or possible child, for the simple reason that abortion does not accomplish such a task.[93]

---

[93] In certain extreme cases, this idea may be warranted (if the disease is so grave that abortion may be said to save the child from a life worse than not coming into being in the

If the goal of the screening is instead taken to be improvement of *population health* (by reducing the incidence of the birth of unhealthy children), abortion can indeed be seen as a treatment (although the "patient" treated is then not any individual, but society or the population as a whole). This is an example of the tension between ordinary individualistic health care goals and those of public health pointed out in Section 3.1.1 and elsewhere.

However, as mentioned earlier, prenatal screening can also be seen as aiming at the goal of promoting the reproductive autonomy of the pregnant woman or couple. In that light, the idea of an opportunity to terminate the pregnancy when the expected outcome does not conform to the plans and wishes of the woman or the couple being suitable as a treatment may look less unwarranted. It should be observed, though, that this goal (of promoting the reproductive autonomy of prospective parents) is in drastic latent conflict with the public health oriented goal. This since, *abstaining from abortion* may just as well be seen as a treatment aiming at the promotion of reproductive autonomy in the face of a positive test-result (depending on the plans and wants of the woman or couple), and that does not square very well with the idea of reducing the incidence of the birth of unhealthy children. What we see, then, is that promotion of reproductive autonomy supports the idea not of *abortion* as a treatment, but of the *option* of abortion as a treatment. This in contrast to the public health oriented goal, from the perspective of which a mere option brings the risk of prospective parents making "the wrong" choices in terms of population health.

There is, thus, a severe potential value conflict built into the idea of prenatal screening with abortion as the treatment on offer in case of a positive test result – a conflict that has to be dealt with if a consistent ethical position is to be reached. Simply put, a choice has to be made between having population health or reproductive autonomy as the guiding value of prenatal screening. As was seen in Section 3.1.1, if the latter is chosen, very few (if any) prenatal screening programmes seem to be ethically justified. Adopting the former value, in contrast, would seem to warrant not only screening but, in fact, would license coercive measures in the reproductive area that most of us would have a very hard time to find acceptable.

---

first place). However, those cases are hardly fitting for screening for a number of other reasons. See Munthe (1996), for more about this.

3.3.2 Counselling as a Treatment

The idea of abortion as a follow-up procedure in a screening context comes out of a situation where the conditions screened for cannot themselves be treated. In other cases, instead, the condition in question might be treated, just not with any medical procedure. As mentioned earlier, counselling with regard to the possibility of affecting life-style factors thought to influence the future occurrence of symptoms may be held out as a viable suggestion for a treatment in the screening context. However, counselling can be suggested as a treatment also on other grounds. For instance, in programmes run mainly to ensure that many people undergo regular general developmental and/or health checkups, counselling aiming to help those who are found to have no problems or to be affected merely by conditions that may become more serious risk factors in the future, counselling is often offered as a purely preventive treatment. Earlier, we have particularly mentioned child and adolescent screening as an example where this approach is practiced, but also less "pure" screening programmes of this sort tend to be similarly organised in this respect – e.g. checkup programmes targeting the elderly or fertile women.[94] This latter use of counselling as a purely preventive treatment for healthy people to stay healthy seems to imply no particular ethical controversy besides those issues connected to counselling in general discussed in Chapter 4. The idea of using counselling as a treatment in the case of positive test results, however, is more controversial as a ground for justifying a screening programme. Nevertheless, there are reasons in favour of such an approach and these are important to consider, since they seem to imply quite different standards for what the counselling may and should involve.

If the view of prenatal screening as serving mainly the procreative autonomy of prospective parents is accepted, counselling aiming at helping them to use the information acquired for the realisation of their own plans may be accepted as a form of treatment that can help to justify screening. However, if the goal of prenatal screening is instead taken to be improvement of the health of the foetus (or future child), counselling can hardly suffice for justifying a screening programme. If the goal of the programme is rather

---

[94] These programmes are less pure than child and adolescent screening in several respects. They target a more restricted sub-population and/or involve health checkups only in certain selected respects rather than the entire developmental or health spectrum. Moreover, since the target population consists of adults, these programmes usually apply less of a proactive and more of an opt-in approach from health care towards the individual.

that of promoting *population* health (by decreasing the statistical incidence of babies with inborn disorders and diseases), counselling may again be accepted as a treatment that could justify screening. However, in that case, the counselling in question would be a *very* different kind of operation than in the autonomy case. For, aiming at promoting population health, the point of the counselling would have to be to try to influence people to choose abortion in case of a positive diagnosis (while, in the autonomy case, the point is to help people make up their own minds).[95]

This rather perplexing situation with regard to prenatal screening can be used to throw some light on other cases, where the suggested follow-up procedure has been counselling, most often aiming at informing people about alternative life-styles that may change their future health risks. The crucial question to ask in these cases, as we have seen, is what goal that may justify such screening programmes in the first place, whether or not this goal supports counselling as a treatment that may justify the programme, and what form the counselling must take in order to be so supported.

As an illustrative example, consider the idea of neonatal, child or adolescent screening for Alpha-1 Antitrypsin Deficiency (ATD), a condition that increases the risk of chronic lung disorders due to exposure to environmental factors such as smoke or dust. This idea was originally proposed with the aim of having counselling about life-style measures that may reduce associated risks (foremost about avoiding parental smoking in the home environment) as the treatment on offer in the programme. In the south of Sweden, such a programme targeting newborns and children was run during 1972–1974, but was eventually shut down after criticism due to negative psycho-social effects, mainly with regard to stigmatisation problems connected to the parent-child relationship. Later follow-up studies (5–7 years) revealed these effects to be both considerable and lasting. At the same time, the expected benefits of the programme in terms of reduced parental smoking did in fact not occur. Rather, after a positive test-result and counselling, parents on the whole smoked significantly *more* than they used to before entering the screening programme.[96]

Despite all of this, it has recently been proposed in rather strong terms that ATD screening should be reintroduced, primarily due to a notable reduction

---

[95] Of course, such a form of counselling would be highly questionable on the basis of the health care ethical requirement to *respect* autonomy. At the same time, we have seen that respect for autonomy is difficult to hold out as an absolute value in the sort of public health ethical context that underlies the idea of having prenatal screening for the sake of population health.
[96] Gustavson (1989).

3.3 Treatments

of smokers among the screened *children*, when followed up at age 18–20.[97] The suggestion has been that such a screening programme should target pre-adolescents:

> As smoking habits are often established in early adolescence, a voluntary screening programme offered to families with a pre-adolescent child would be advisable from a somatic point of view. At that age, the psychological stigmatization from labelling the child as deviant and the possibility of negatively influencing the parent –child relationship should be considerably reduced.[98]

No empirical support has been presented in support of the statement about reduced social risk. Moreover, the problems particular to child and adolescent screening (see Section 3.1.3) are not addressed. For instance, if the screening is supposed to be voluntary, who should consent: the parents, the children, or both? And what should be done in the case of conflicts between parents or between parents and children? In our view, it is imperative that all of these question marks are straightened out before screening can be considered as a serious suggestion. Also, we suggest that when screening is suggested on such shaky ground as in the just related case, this may quite generally be an indication of the strong pull that screening programmes seem to have on many medical researchers and professionals being at work (see Chapter 1).

This example is of interest to illuminate the problems of the general idea of screening with counselling as the only sort of treatment on offer as a response to positive test results whenever the main objective of the programme is to influence health, be it on the individual or the population level. The remaining objective to consider would then be the idea of promoting autonomy. However, both the stigmatisation risks illustrated by the ATD-case and the general reasons to be sceptic about the potential of screening programmes to actually further autonomy held out earlier (see, e.g., Section 3.1.1), would seem to tell against also that idea. In conclusion, therefore, while counselling is indeed a very important ingredient in any screening programme (a point further elaborated on in the next chapter), and may be a central as a response to those who test *negatively*, the prospect of justifying such programmes in virtue of a follow-up procedure to *positive* test results consisting of *mere* counselling does not look very bright.

---

[97] Thelin et al. (1996). It should be noted, from a methodological point of view, that the study included very few individuals (50 diagnosed with ATD and 48 in the control group).
[98] Thelin et al. (1996), p. 1211.

3.4 Summary

Screening programmes have to be evaluated in terms of the goals they try to achieve and the overall balance of their costs and benefits compared to alternative measures. Several more specific factors have to be considered in order to operationalise goals, benefits, and costs. In this chapter, we have been identifying and investigating such factors. They include properties of the condition in question, the tests and analytical methods applied, and the follow-up procedures offered as response to negative as well as positive test results. Since different conditions have their onset in different ages, there is also the related question of when in the patients' lives the screening should be applied – during the prenatal, neonatal, childhood or adult stage.

First, the conditions screened for must be an important health problem, which means that their prevalence in the general population is sufficiently high and that the condition is sufficiently serious for the affected individual. If a condition is very serious, prevalence need not be as high and vice versa. The actual practice of screening oscillates between giving prevalence the decisive weight and holding out severity in the single case as their primary rationale. However, since different screening programmes can have different goals depending on who is the target of the programme, this means that the assessment of the ethics of such a programme needs to proceed from different outsets depending on the target population. For instance, reproductive autonomy is foremost used as a goal for prenatal screening, while other screening programmes with other targets aim for other goals. Thus, the goals can affect what conditions should be considered serious enough for screening.

However, even if one grants that reproductive autonomy is an important goal for prenatal diagnosis, it is a highly questionable argument in favour of organising this practice in the form of screening programmes. This in view of the obvious risk of such an organisation of the practice to express discriminatory messages about disabled people, to undercut patient autonomy, and noting that the chief aim of using the methods currently applied in prenatal screening (to reduce the amount of invasive prenatal tests) can be attained without organising the practice as a screening.

As regards neonatal screening, all developed countries have more or less ambitious neonatal screening programmes. The diseases screened for in these programmes have traditionally shared some common characteristics: they can be detected at this early age, the condition most likely had not been detected if not screened for and early measures are available as well as necessary for preventing or ameliorating the diseases in question. Moreover, the

3.4 Summary

tests for the diseases are very reliable and safe, and the diseases themselves are very serious. However, depending on how the criteria and goals are interpreted in detail, as well as differences with regard to socio-economic and political structure, the policy and practical recommendation for neonatal screening have come to vary considerably between countries. In general, USA is still more screening-friendly than Europe and it is also in the US context where we find examples of denying parents any option to withdraw their babies from screening (although the pressure to participate is generally quite distinct in all neonatal programme) as well as ethically highly dubious examples of applying such options to parents only in the case of religious objections. We have highlighted how the expansion of neonatal screening taking place in recent years makes the idea of giving parents no or very little room for withdrawing their babies from neonatal programmes less defensible from an ethical point of view, but also argued that if parents are given the opportunity to consent, this consent should be *informed* (thus based on appropriate counselling). Difficulties to achieve this in a cost-effective way may, we have claimed, provide a substantial ethical reason against the more ambitious expansions of neonatal screening. At the same time, the general issue of parental informed consent in the screening of children is indeed complex and therefore considered further in other parts of this book.

All screening programmes targeting children and adolescents give rise to some complicated issues. Some parents may, for example, be unwilling to have their children entering the programme – while this used to be a minor phenomenon in neonatal screening, the above mentioned expansions of such programmes has made it more likely that also they will be more affected by such an issue. Children may, moreover, be harmed by screening programmes through the phenomenon of stigmatisation, at the same time as children may be more or less autonomous themselves. In relation to adolescent and adult screening, we also emphasised the need for ethical assessments of screening programmes to consider a wider socio-political context. This is so, partly due to the fact that the information provided by the tests used in screening programmes may be of interest not only for health care and the individual, but also for various third parties, e.g. relatives, insurance companies, and employers.

As regards the properties of the methods for investigation, testing and analysis applied, we addressed three aspects with regard to this: safety, validity and predictive value. One of the most important conclusions from that analysis was that in order to determine whether to opt for higher sensitivity or higher specificity, a closer examination is needed of various other variables, e.g. the conditions targeted, the consequences for the individuals tested as well as for society and health care in general, and the presence of treatments. Another important conclusion was that, as a rule, the less predictive value, the stronger the reason to abstain to use the test in a screening program.

The difference between ordinary medical testing and testing done within a screening programme also has consequences as regards treatment, e.g. that the general responsibility of health care to be able to offer an acceptable treatment at the outset is stronger in the screening case than in the ordinary health care situation. Moreover, we argued that a choice has to be made between having population health or reproductive autonomy as the guiding value of prenatal screening, that it is more plausible to choose the latter, but that this speaks against organising prenatal testing as screening. Similarly, the prospect of justifying any type of screening programmes in virtue of a follow-up procedure in the case of positive test-results consisting of mere counselling does not look very bright. At the same time, counselling may often be a very valuable response to *negative* test-results, especially in broad programmes consisting of general developmental and/or medical checkups where the opportunity of helping healthy individuals to stay healthy adds a strong rationale.

Chapter 4
Screening – How?

Besides the actual testing and analysis of samples, application of follow-up procedures, et cetera, screening programmes also involve the processes of contacting people for recruitment to the programme, informing them about the procedures prior to testing, as well as about the result of the test afterwards, counselling about possible follow up-procedures, and help with coping with the reactions to the test result. These features of screening programmes give rise to a host of questions of *how* screening programmes should be designed and conducted in these respects. Plausibly, all programmes should involve all of the tasks mentioned to some extent. But how much? And in what way? If ill-designed, the programme may end up not promoting the values it could have promoted and producing negative side-effects it could have avoided. So, even if defensible in terms of the condition targeted, the testing method utilised and the treatments available, a programme may still be open to serious criticism if organised in an inferior way. This has been underscored, e.g., in research on the new risk assessment methods in prenatal screening discussed in Section 3.2.1.[1]

We will not address all possible questions regarding this, but rather concentrate on some of the most debated and important ones. First, there are issues regarding how screening programmes approach and subsequently handle people. With regard to this, we will focus on the much debated and related questions of informed consent (Section 4.1) and counselling (Section 4.2). Second, we will address some "large scale" issues regarding funding: if, and then how, participation should be encouraged by the use of, e.g., financial incentives (Section 4.3). An important general point to keep in mind in the following is this: just like the values or goals of screening have important implications for what screening programmes that should or should

---

[1] Saltvedt (2005).

not be offered, they likewise have implications for *how* such programmes should be offered.

4.1 Informed Consent

Even if a large uptake traditionally has been considered desirable in screening, today there is a general agreement that participation in programmes targeting adults should be voluntary and that the patient has a right to be informed about the procedure and the possible effects of accepting or rejecting it.[2] In other words, *informed consent* is considered to be an important feature of an ethically sound screening programme. As we have seen in the preceding chapter, when the programme is instead targeting neonates, children or adolescents, there is less of consensus as to whether and, if so, how and to whom informed consent is to be applied.

However, the consensus on informed consent in the case of adult screening has not always been a given. In the 1970s, mandatory screening was defended[3] and the attitude towards voluntariness and informed consent in connection to the prenatal screening programmes that begun to be launched in this period was rather equivocal at the time.[4] It is perhaps symptomatic that such ideas have evaporated against the background of the increased emphasis on autonomy as a central norm and value of health care. As mentioned in Section 2.2 above, the basis for rules of informed consent is the rule of respecting autonomy. Thus, in the context of discussing informed consent, considerations of autonomy take the centre stage.

Respecting autonomy is incompatible with coercion and manipulation and informed consent is a way to avoid exactly that. Being informed about a procedure and its effects is necessary in order to ensure that one makes a decision in accordance with one's own values and wishes. Withholding or distorting information relevant to the decision of a patient can lead her to make another decision than she would have made in the light of that information. Thus, such actions amount to manipulation of a sort difficult to square with the ideal of respecting autonomy. Furthermore, in order for an acceptance to go through with testing to be one's own autonomous decision, voluntary consent is necessary. Thus, deliberately ignoring to obtain the consent of the patient amounts to coercion, as does subjecting her to

---

[2] See e.g. ESHG (2003), p. 56, point (12) and (23).
[3] Hoedemaekers (1999), p. 216.
[4] Munthe (1996).

4.1 Informed Consent

strong pressures or threats in order to secure consent. For instance, the standard rule of clinical research ethics, that the offer of participating in a study must in no way be connected to people's access to health care, seems to apply equally to screening.

However, how the procedure of informed consent should be designed in order to ensure that autonomy is respected is a tricky question. For one thing, relating to the "consent"-part, it is unclear what amount of pressure to make one decision rather than another that amounts to an illegitimate infringement of the patient's autonomy.[5] Obviously, threats of sanctions, such as fines or lack of future health care if one chooses to abstain from entering a screening programme, clearly amount to coercion. However, more subtle forms of pressures, like more or less open discontent with certain choices, may also be seen as problematic from the point of view of autonomy, since it makes some alternatives "harder" to choose for the patient, even if such pressure does not clearly constitute coercion. In fact, as argued earlier, the very feature of a testing-offer to come in the form of a screening in itself always brings some pressure to go through with testing (see Section 2.2). Another aspect is if the offer is connected to some sort of positive incentive, such as extra access to medical consultation, investigation, et cetera, over and above that offered in the case of a positive test-result.

In addition to the clinical handling of presenting the offer to enter a screening programme, societal attitudes can invoke similar pressures, e.g., with regard to prenatal screening (see Sections 2.2, 3.1.1, and 5.1). Likewise, the arrangement of societal institutions can make some decisions more difficult to make in practice. For instance, the more burdensome to raise and care for children with certain impairments and disabilities, the more the "right" to abstain from prenatal testing will be formal than real. Of course, how burdensome this is in practice is determined both by societal attitudes and by institutions. Indeed, the degree to which many impairments are disabling is determined by society, at least to some degree.[6] Thus, it can be argued that the ideal of requiring voluntary consent in fact has far-reaching consequences for how we should arrange society, e.g. with regard to the opportunities and support for people with disabilities.[7]

Relating to the information-part of informed consent, the default position seems to be that "everything" regarding testing and its possible effects should be disclosed to the patient. It seems insufficient only to disclose

---

[5] See Wilkinson (2003), chapter 6, for a general discussion of this question.
[6] Parens and Asch (2000).
[7] Munthe et al. (1998).

information that the patient explicitly regards as relevant at the initial stage. This since it is plausible to assume the uninformed patient to be in a bad position when it comes to knowing what *type* of information that might be of interest for her. For example, a person considering whether to accept an offer of a carrier detection test for cystic fibrosis may be unaware of the increasing possibilities to treat the symptoms of the disease, information not unlikely to be considered as relevant for the decision if it was to be disclosed. So information must include the natural history, basis, and possible symptoms of the disease in question, what the test can show, how a positive and negative result should be interpreted, and what follow up procedures that are available depending on the result.

Furthermore, it is of essence that information is not only confined to medical facts and medical consequences of alternative decisions. This is so, since a test-result can affect many aspects of a person's life. For instance, in order to make a well-founded decision of whether or not to go through with some carrier screening or prenatal test, it is crucial to know what life may be like if one gives birth to a child with the disorder in question. Once again, this is to a large degree determined by societal factors. Another example of something that can be of relevance to the decision to go through with testing is one's possibility to purchase personal life- or health insurance, or one's chances of securing or holding a job (see Section 3.1.4). In general, the societal and psychological consequences of receiving a result can be as important as the medical one's.[8]

However, the idea of disclosing "everything" and thus not withholding any medical or other information that may be of relevance to the patient is the subject of much discussion.[9] First, this rule should not be taken to mean that information should be *enforced* on the patient without her consent. The right *not* to know can also be defended with reference to considerations of autonomy.[10] This is so, since forcing information upon someone against her will could also interfere with her important plans and values. As a consequence, the health care professionals initially have to obtain information from the patient what she would *not* like to find out about. In effect, from an autonomy perspective, the programme needs to be organised in a way that makes this essential step possible.

Second, large or complicated bundles of information may *reduce* autonomy, since more confusion than clarification may be the result. This can

---

[8] This is increasingly recognized in the literature on genetic counselling (see e.g. Platt Walker (1998); Munthe (1999), p. 83; and Juth (2005), pp. 84–85).

[9] See e.g. Beauchamp and Childress (2001), pp. 83–88; Hoedemaekers (1999), p. 219; and Juth (2005), pp. 92–98.

[10] Häyry and Takala (2000); and Juth (2005), chapter 5.

either be seen as an argument in favour of withholding information or an argument in favour of implementing procedures that facilitate the processing of information, such as counselling. If respect for autonomy should be given a central role, the latter point of view is the more plausible one (see next subsection). However, *very* much information may be hard to process anyway. Due to this, programmes that simultaneously screen for a large number of disorders may be problematic indeed from the point of view of autonomy, and have sometimes been seriously questioned by professional organisations, e.g., within genetic medicine.[11] Similar problems are created by multi-stage methods of calculating risk, such as those employed in the new risk-assessment methods in prenatal diagnosis using ultrasound and serum-tests in addition to maternal age (see Sections 3.1.1 and 3.2.1). At the same time, broad screening programmes consisting mainly of ordinary medical checkups (where risk indicators are clear and treatments readily available) may appear rather unproblematic, since they offer clear benefits in terms of both individual and population health (regardless of if the test result is positive or negative).

A further factor to consider is that, in many programmes, unexpected information may be revealed. This has been highlighted, particularly, regarding genetic tests, where the example of non-paternity provides a classic example. The rich palette of possible secondary findings in expanded neonatal screening programmes (see Section 3.1.2.3) is a more recent case. It is not clear from the point of view of health, psychological well-being, or autonomy whether and, if so, how such information should be disclosed.[12] This problem can be sidestepped to some extent by finding out beforehand what possible information from a test that the patient would like to remain ignorant about. However, if also the professionals have failed to foresee the possibility of some resulting information, this manoeuvre will be unavailable. This indicates, if nothing else, that screening programmes need to have built into their organisation a readiness to handle the phenomenon of unexpected or secondary information and, perhaps, that this readiness needs to include disclosing such possibilities to people in the process of collecting informed consent.

Now, the goal of promoting health can obviously conflict with the rule of respecting autonomy, since individuals need not consent to testing and treatment beneficial to their own health or that of the population. Thus, despite an increasing allegiance to the norm of respecting autonomy, and consequently to informed consent also in screening programmes, it might be

---

[11] ESHG (2003), p. 56, point (27).
[12] Juth (2005), pp. 92–98.

argued that, in some cases, informed consent should be a secondary concern in screening programmes after all. In the preceding chapter, we encountered such reasons connecting to the fact that members of the target population may not be fully decision competent, but the argument may also be advanced as regards competent adults. Judging from the case of communicable disease discussed in Chapter 2, acceptance of such a view might seem more tempting if the goal of screening is foremost to promote population health, but we have also seen strong basis for questioning such thinking in other areas, e.g., prenatal screening.

However, the basis for respecting autonomy in screening need not be the idea that such respect is a fundamental moral duty. This norm can also be based on more instrumental concerns, such as preserving trust in health care as well as public health institutions and arguments to the effect that no one has a greater knowledge of and interest in her well-being than the individual herself.[13] Nonetheless, if such instrumental considerations are the basis for respecting autonomy and obtaining informed consent, such respect is only warranted to the extent that it is, in fact, an efficient means to these basic ends.[14] Since this may vary from case to case, the instrumental line of reasoning therefore seems best suited for defending the informed consent standard as a general rule at an institutional level, acknowledging that (just as in the case of most institutional rules) there may be singular cases where application of the rule cannot be defended on the basis on which the rule itself is defended.[15] Moreover, even those who argue that respect for autonomy is of importance in itself (regardless of beneficial overall consequences of adhering to such a standard), are inclined to say that the reasons for such respect can be overridden by other considerations if these are weighty enough (e.g. in terms of health promotion or avoidance of harm).[16] Again, in practice, an institutional perspective seems to be the one where decisions regarding which such proposed overriding circumstances should prevail have to be made. So, even if disclosing all relevant information is adopted as the default position in health care also in the case of screening, the questions about the basis for and the limits of legitimate informed consent (e.g. the tricky question about parental informed consent in expanded neonatal screening programmes presented in Section 3.1.2.2) are far from settled,

---

[13] See Tännsjö (1999), where a radical ideal of anti-paternalism in health care is defended on the basis of this kind of considerations.
[14] Nijsingh (2007).
[15] Sandman and Munthe (2009).
[16] See Beauchamp and Childress (2001), chapters 3–5. The position that respect for autonomy should only be prima facie and not absolute, regardless of basis, is argued in Juth (2005), pp. 205–206.

and to settle them would seem to require an ethics perspective that lifts itself above the individual cases and takes an institutional view of screening programmes, health care, public health efforts and their overarching societal functions and effects. An outline of such an approach and what it may imply (e.g. regarding parental consent in neonatal, child and adolescent screening and prenatal screening) will be presented in the final chapter.

4.2 Counselling

In order for individuals to benefit from the medical information resulting from a screening programme, *what* information is disclosed and *how* it is disclosed is of great importance. As pointed out in Section 2.1, this is important from the point of view of health, since understanding the information properly is necessary in order to undertake appropriate health care measures and to avoid unnecessary anxiety. Moreover, as we saw in Section 2.2 and the previous subsection, the manner and form of disclosure is also important from the point of view of autonomy, since understanding information properly is necessary in order to be able to make decisions in accordance with one's plans. However, it is worth underscoring that failure of or difficulties in disclosing information in an appropriate manner is far from being a problem only from an autonomy perspective. Disclosure is of importance both for having participants in screening programmes more prepared for the results and what they may imply in terms of follow-up procedures and necessary adjustments in one's daily life, as well as for having those tested understand the significance and practical implications of the eventual test results and thus being better equipped to participate in whatever follow-up procedures are subsequently applied. This is just as much of importance from a pure health perspective (both individual and on a population level) as it is on the basis of autonomy considerations. In effect, the organisation and practices of a screening programme in these respects may very well function as a reason against having this programme on the basis of health considerations alone.

Counselling is a practice that aims at designing the situation of disclosure so that it is conducive to the health, well-being and autonomy of the individual. It is, in the words of Fraser's famous characterisation of genetic counselling, a "communication process which deals with the human problems associated with the occurrence or risk of occurrence of a genetic disorder."[17] The fact that this description concerns *genetic* counselling in

---

[17] Fraser (1974), p. 637. Adopted by the American Society of Human Genetics in 1975, this conception of genetic counselling has since then become standard.

particular is symptomatic of the situation of this particular area of medicine, where many conditions can be described and predicted, but not treated to any greater extent, which creates special challenges for counselling. The attention to counselling within clinical genetics is partly due also to genetic information having been considered as *special* in various ways, e.g. by being revealing of others (blood relatives), often very complex (see Section 3.2), and especially sensitive. However, since it can be argued that other medical information is special in relevantly similar ways, e.g. information about being a carrier of HIV[18] or the information that one bears signs of an increased risk of (non-hereditary) cancer, there is no reason in principle not to have ambitious counselling procedures for such information too. As a matter of fact, counselling seems to be a central aspect of many screening programmes for children and adults where these programmes create the opportunity to help healthy people stay healthy. Also, in general medicine, counselling is a necessary means to having patients and their close ones understand the implications of various conditions and applied treatments for their daily lives, so that they can make the adjustments necessary for the treatments to be as effective as they can be. Nonetheless, genetics is the area where, over time, the notion and idea of medical counselling has been most developed and debated from an ethical point of view.[19] Because of this, we will initially focus on genetic counselling too, adding as we go along how the various points raised in the genetic counselling context seems to fit health care more generally.[20] On that basis, we will then address how the lessons learned may apply more generally to the disclosure of information in the health context, connecting these ideas to a common underlying rationale by referring to so-called shared decision making.

4.2.1 Genetic Counselling as a Template

Genetic counselling can be more carefully characterised by a number of components that become intelligible in light of promoting the values of autonomy and health. First, Fraser's influential characterisation says that genetic counselling is a "communication process". This indicates that it is a

---

[18] Holm (1999).

[19] For passages on the history of genetic counselling, see Juth (2005); and Platt Walker (1998).

[20] That ethically defensible testing, whether done in screening programs or not, should include counselling is almost unanimously accepted. See e.g. ESHG (2003), p. 56; and WHO (1998), p. 9.

4.2 Counselling

question of interaction between the counsellor and counselee, rather than the doctor telling the patient what the medical condition is and what she should do about it. This is well in line with the autonomy-inspired idea that it is the values and concerns of the patient that should govern genetic counselling (nowadays predominant in that field). However, it just as well connects to the general experience in medical practice that, in order for patients to be able to adhere to health advice, they often need to engage in some exchange with health professionals – implying that the latter need to tailor their counselling approach appropriately in order to have the desired health-related effects. The mention of a *process* thus indicates that counselling "ideally takes place over a period of time so that the client can gradually assimilate complex and potentially distressing information."[21]

Second, Fraser mentions that the counsellor should be "one or more appropriately trained person".[22] The idea is that the complex information that is communicated and the sensitivity of the process of doing so requires not only knowledge of genetics, but also pedagogic skills to ensure comprehension or *understanding* (rather than just *disclosing* information), caring skills of helping the patient to cope with psychological and social consequences in order to accomplish the goals, and so on. Acquiring such a blend of skills will presumably require some training. Also this notion seems to fit the more general case of patient-health care interactions, where the medical knowledge of the professionals needs to be complemented by further skills in order to have a successful communication process. In more mundane health care situations, these skills may be assumed not to require so much of special training. However, when the situation has similarities with the case of genetics in terms of the complexity of the information and the various troubles that may occur when people try to relate this to their life, the practice of counselling becomes more demanding also in other areas of medicine.

Fraser also makes a rough division between two parts of the counselling process: pre-testing and post-testing.[23] Pre-testing refers both to the process before the decision whether or not to take a test and communication occurring in the actual testing situation. Pre-testing involves *information gathering* on the part of the health care professional,[24] involving information

---

[21] Platt Walker (1998), p. 5.
[22] Fraser (1974), p. 637.
[23] There are genetic analyses made without molecular or other biochemical testing, most notably by making a pedigree of a family history of disease. For those cases, it is more appropriate to talk about pre- and post-analysis.
[24] Platt Walker (1998), p. 9.

about the patient of a "medical" kind, as well as information of more "psychological" and "social" kind. The idea is that, in order to help the patient with achieving her aims, coping with result of the test and to achieve optimal health outcomes, the counsellor needs to investigate the motivation and general situation of the patient, what she expects to learn from testing, how she is likely to respond to different test results, what her fears and hopes are, and so on. The patient, in response, is informed about possible outcomes in relevant terms – not merely regarding risk for disease for herself and her family and available preventive measures and treatments – but also in terms of psychological reactions and social and economic consequences. The idea is that the patient should also be informed about means of support to cope with such consequences. Also these ideas seem to fit well with practices in general medicine, not least the classic practice of *anamnesis* as the starting point of any session of medical consultation.

If testing is subsequently done, then enters the stage of post-testing. Post-testing counselling involves *information giving* of risk for disease,[25] where, of course, it is material that the process leads to an adequate understanding of the test-result. Information giving moreover actualises questions about follow-up procedures, treatments or general advice in relation to what the patient faces as a result of the test. But post-testing may also include psychological counselling and other means of support from the counsellor in order to handle the possible psychological effects of learning the result, such as distress, anxiety, guilt and strained family relations. Of particular interest is to mention that this by no means concerns only cases where the result is positive. It is well-known from the practice of genetic testing that negative results may create feelings of poor health (if the person is identified as a healthy carrier of a "disease gene") or so-called survivor's guilt towards relatives who have tested positively.[26] In the more general health care context, when some investigation has been undertaken, information giving is, of course, as obvious as in the genetic case. Moreover, also in other areas of medicine, coping with various psychological reactions to the test result may be needed.[27] In contrast to the genetic case, finally, investigations revealing no special causes for worry may actualise counselling about how the patient may act as to sustain his or her health status.

Genetic counselling has developed an ethos, containing particular norms or principles, the compliance to which are thought to promote the overall

---

[25] Platt Walker (1998), p. 10.
[26] Sobel and Cowan (2000).
[27] For instance, the increased risk of suicide due to a cancer diagnosis (see Section 5.4).

4.2 Counselling

values strived for in this branch of medical practice. These include minding about the psychosocial and affective dimensions of the patient, as well as the upholding of confidentiality and protection of privacy.[28] However, the most salient and debated of the norms of genetic counselling is that of *non-directiveness*.

There are controversies about the meaning of non-directiveness and doubts have been expressed whether it is a plausible, or even feasible, norm for (genetic) counselling at all.[29] However, there seems to be consensus on the view that the point of non-directiveness is primarily related to concerns about the patient's autonomy. The idea is not only that the decision of the patient should be respected, i.e. the traditional one in biomedical ethics, but also that "ND [non-directiveness] describes procedures aimed at promoting the autonomy and self-directedness of the client."[30]

However, there are controversies about what this implies in practice. Sometimes the norm of non-directiveness has been taken to imply that the professional should abstain from attempts to influence the decision of the patient.[31] However, avoiding any influence over the patient seems to be impossible, a fact that is widely acknowledged.[32] Just by presenting some information, a patient's decision can be affected. For instance, revealing some piece of information about preventive measures, previously unknown to the patient is not unlikely to influence her decision, e.g. to use that measure. There are other more subtle ways of influencing, e.g. by the way information is presented (through choice of words, body language, etc). So non-directiveness in the sense of avoiding *any* influences *whatsoever* on the patient's decision is not a plausible ideal for counselling – be it in the realm of clinical genetics or health care in general.

The common interpretation of non-directive counselling as counselling without *advising* or *recommending* the patient what to do[33] is not very plausible either, since giving advice can in fact promote the patient's autonomy. For instance, serving the patient's autonomy may include recommending her to see a psychologist in order to get a better grasp of what she actually is trying to achieve.[34] Moreover, the advice to see a psychologist can help a patient in coping with paralysing emotional distress, helping her to

---

[28] Juth (2005), pp. 79–97.
[29] Juth (2005), pp. 85–87.
[30] Kessler (1997), p. 166. See also Munthe (1999), p. 82.
[31] ten Have (2000).
[32] See Juth (2005), p. 87 for a discussion and further references regarding this.
[33] Juth (2005), p. 89.
[34] Munthe (1999), p. 85.

make her own decisions and to live in accordance with them. So abstaining from advice should not be an integral part of non-directiveness.[35] Rather, it is a certain *kind* of advice that one should have in mind when being critical against advice-giving from the counsellor, namely advice aimed at making the patient decide in a certain way *regardless of the desires, aims, and values of that patient.* Problems thus arise when the counsellor brings her own normative agenda (or that of some institution she represents) to counselling sessions and lets that agenda overshadow what the patient finds important. When this happens, all kinds of influence, not just advice, become problematic. Overemphasising the risk of certain alternatives while playing down risks of others, failure to bring out certain information about possible consequences (including psychosocial), manner of behaviour, like expressing, however subtle, discontent when patients want to develop certain aspects, are all examples of ways of illegitimately influencing the decision of the patient.

As mentioned, this stance to counselling developed within the clinical genetic field often assumes autonomy to be a supreme underlying value. While this is understandable due to the mentioned situation that has long affected genetic medicine of lacking any cures or other effective treatments for most of the diseases dealt with, it makes for some difficulty to transfer this idea into health care in general. In fact, this difficulty appears already if we consider the few cases of (primarily) genetic disease for which there in fact exist effective treatments – where it would be quite absurd to suggest that the future health and well-being of the patient is of no concern. Rather, it seems to us, to the extent that there are indeed unequivocal prospects of protecting and promoting the health and well-being of patients, this should be a concern for counsellors just as much as that of promoting autonomy.

Thus, the conclusion must be that the sound core of non-directiveness that may be applied to counselling in health care generally is about abstaining from influence that may be detrimental to the autonomy, health or well-being of patients, but embracing other kinds. Roughly, all influence should be exercised with an eye to assisting the person in becoming more autonomous – the desires, aims, and values of the patient must therefore guide the counselling.[36] But this does not exclude that the counsellor or health care influence patients to consider options and reasons on the basis of

---

[35] This is also increasingly recognized in the discussion about genetic counselling. See Shiloh (1996).
[36] Kessler (1997), pp. 169–170.

considerations of health and well-being. When it comes to the practical question of how to achieve this kind of non-directiveness, training in counselling skills and practical experience of counselling is essential.

4.2.2 Expansion: Shared Decision Making

The idea of counselling developed within the genetic context as characterised by dialogue for the communication and gathering of information of both of the involved parties – the patient and the health care professional – with the aim of promoting patient autonomy, corresponds rather well with the more generalised ideal of patient consultation and health care planning developed under the heading of *shared decision making*. Especially in its more developed variants this approach rests on notions of how information disclosure, forms of communication and the balance between describing options and giving clear advice can be optimised in light of aims to promote autonomy and health.[37] Shared decision making in such a form thus appear to be very similar to the more developed versions of the notion of non-directive genetic counselling.

At the same time, the notion of shared decision making is not restricted to the area of genetic medicine, but is intended as a model for consultation, planning and decision making in all sorts of health care. Hence, this concept can be used as a theoretical tool for making the lessons learned regarding counselling in the particular genetic context and how it seems to relate to other areas more explicit and general. Especially so, since the ambition to consider a wide range of personal and contextual factors that may affect how health information and health care measures actually affect and influence people held out as fundamental in advanced models of shared decision making is very similar to genetic counselling. In addition, the notion of shared decision making provides a theoretical framework for explaining and assessing how councelling may incorporate attention to both autonomy and health/well-being.[38]

Applied to the case of screening, to conceive of the counselling task in this light may provide help in clarifying how the balancing between health and autonomy considerations may proceed in different stages. A shared decision making process may very well start by having health care approach someone who is unprepared for this with an offer of undergoing some medical investigation or test, if only there are apparent benefits of such an approach

---

[37] Sandman and Munthe (2009, 2010).
[38] Sandman and Munthe (2010).

in terms of health and/or autonomy. This benefit appears when the screening approach is compared to alternative options, e.g., an offering that is organised in a less proactive form than directly approaching unprepared people with explicit requests. The counselling implied at this stage is about securing that people who join the programme are aware of what they are joining. The importance of this can be supported not only by reference to autonomy, but equally well to considerations of health – be it on the individual or population level (see below). The process applied to accomplish this will often have to involve substantial elements of dialogue, since the health care professional often needs to understand what information is relevant to the particular patient, what sort of follow-up procedures she is open to considering, what information she may want not to receive, and what way of communication is suitable to facilitate understanding. This process may involve the giving of advice, as well as designs of the offer that are of a more "opt-in" nature. It also needs to consider the psycho-social needs of the patient, e.g., for adjusting to a situation which she has not anticipated and for which she is not prepared. The role of post-testing counselling in line with the shared decision making approach is very similar; only, in this case, follow-up procedures to the result of the test are included among the options among which the patient needs to make a choice, or choose to ignore entirely.

There are good reasons to assume that the quality of pre-testing counselling will very much determine the prerequisites for successful post-testing counselling. Regarding some screening programmes, this may not appear to require such ambitious efforts, since people can understand quite easily what the point of the programme is, e.g., regarding developmental and health checkup programmes for children, or the traditional forms of neonatal screening. However, in these cases, this is mainly because the general societal organisation around reproductive and paediatric care have prepared them already. Without such preparation, not to make ambitious efforts to have people understand the point and nature of these programmes would be asking for trouble. Thus, a general conclusion regarding all screening programmes is that they should involve an ambitious counselling organisation already at the outset of the programme, making sure that people who do enter have had a good chance of forming an understanding of what it is they are initiating in light of their own aims and values and their own particular social context. Such a set-up can be anticipated to facilitate a smoother post-testing counselling, by forming a reference point for the decisions and reactions that arise as a result of the test.

It should be noted that none of the above *necessarily* implies that the implementation of a screening programme may never involve autonomy reducing or restricting elements. Shared decision making as such does not imply any definite view on how the values of autonomy, individual health and population health should be balanced in cases of conflict[39] (as mentioned before, we will return to such issues in the final chapter). It does imply, however, that people targeted by a screening programme should as a rule be helped to become aware of what they are a part of even in cases where there are reasons to downplay the importance of autonomy. This for the simple practical reason of the importance of, first, being able to have the entire chain of events (from initial approach to the last action taken by the programme organisation *visavis* a person) during a programme run smoothly and effectively and, second, sustainably preserving the public trust in the programme as well as in health care and general public health institutions. It may be further observed that the troubles created by not meeting this need will often lead to elevated societal costs further down the line, not least in ordinary health care or the social services (where the resistance, confusions, misunderstanding and results of misguided choices due to poor counselling at the outset will as a rule have to be dealt with). In cases where autonomy is in fact an important factor, this adds further reasons to the same effect.

This presents a special problem for screening programmes, since the above seems to imply that the counselling organisation needed to be set up will as a rule add considerably to the costs of the programme. This at least places the onus probandi on the party proposing such a programme: benefits must clearly outweigh harms in order for the implementation of a screening programme to be justified, and the measures necessary for securing that so is the case (such as counselling) must not make the programme too costly. As indicated in Section 3.1.2.2, this may mean that programmes with less uncertain benefits but implying rather large counselling needs (such as the more expanded version of neonatal screening) are, as a matter of fact, not very cost-effective at all.

4.3 Funding and Participation

An ethically acceptable screening programme will, thus, most likely be an expensive affair, and this, in turn, may in fact make the programme cost-ineffective. This raises the further issue of how screening programmes should be funded? Normally, it is assumed that individuals themselves should not pay for their costs. There are good reasons for this assumption:

---

[39] Sandman and Munthe (2009).

most likely participation will be affected negatively if people have to pay for themselves. Moreover, it seems unjust that only those who can pay for themselves should be able to benefit from such programmes. The injustice seems extra obvious in the case of screening, since the initiative comes from health care and not from the individual herself: should individuals really have to pay for something they are requested to do?

However, this gives rise to an issue of allocation: To what extent should publicly funded health care spend resources on screening? This question, in turn, awakes a host of difficult questions: Should health care primarily spend its resources on people with identified or clearly experienced health problems or people who may or may come to suffer such problems? How great should the benefits of screening be in comparison to the benefits of alternative health care measures? Which benefits and harms should count? Health only, or also other values, such as autonomy and psychological well-being? How should they be balanced in cases of conflict? What role should financial savings and considerations of efficiency play?[40] In this book, we have addressed some of these questions, e.g. regarding the basic values of screening: which benefits and harms should count and how they should be balanced in cases of conflict. We have argued that there is no reason in principle to only count benefits in terms of physiological health (Chapter 2). For instance, autonomy and psychological well-being are considered values in medicine in general, and we can think of no principled argument for why they should cease to be considered as such in the area of screening. However, as we repeatedly argue in this book, there is often reason to believe that screening is not a very good mean to realise these values. The most important rationale for screening thus remains gains in terms of physiological health (e.g. measurable in terms of mortality and morbidity).

The rest of the questions regard to what extent and in what ways we should have different demands of cost-efficiency as regards screening and other preventive medical efforts as compared to traditional curative and ameliorating individual ones. Although there are some clearly stated opinions regarding this intriguing but difficult issue,[41] surprisingly little has been explicitly said about it. We will not try to answer it either, although we will return to it in the final chapter in general terms (Section 6.5.3). Nonetheless, it is important to point the question out and note that there is some work to be done both in public health ethics and health policy.

---

[40] Cf. Danish Council of Ethics (1999), chapter 5.3.
[41] E.g. Justman (2010) claims that harms should be weighed considerably higher in preventive medicine (see Section 5.3).

A related question is how to ensure participation in an offered screening programme. This question is relevant when the possible benefit is in terms of public health: in order for a screening programme to improve *public health*, i.e., in order to make a notable difference in terms of increasing life expectancy, and decreasing morbidity and mortality at a *population level*, participation of a sufficient number of individuals is necessary.

As already mentioned, reimbursement of monetary costs for the individual most likely increases incitement to participate. Another strategy is public education about the benefits of the screening programme. However, this strategy may backfire if the benefits are dubious or if people are harmed in various ways by participating, at least in societies with free media. So benefits have to be obvious not only in order for a screening programme to be ethically acceptable, but also in order to be accepted by the public, at least in the long run.

Yet another way to increase participation is to organise the setting in which it is offered in a certain way. For instance, one can offer screening in connection to other health care measures where uptake is large, as is the case of prenatal and neonatal screening. Moreover, explicit verbal offers are more likely to encourage participation than is, e.g., handing out a written form or just posting information and an invitation on a website.

Of course, there is always the possibility of individual persuasion in order to increase participation. However, not only is this problematic from the point of view of autonomy (see Sections 2.2 and 4.1), but also could undermine the legitimacy of the programme, at least if persuasion is widespread. The more persuasive features contained in the offer, the more people have reason to ask themselves: "If they need to muster such a campaign to get people on board, how beneficial can it be, really?" However, as we will see, some screening programmes are very persuasive and have been so for quite some time and still enjoy (unrealistically) high appreciation in the eyes of the public.[42] The greatest problem with these programmes is not lack of legitimacy, but failure to fulfil reasonable standards of informed consent.

4.4 Summary

Screening programmes involve the processes of contacting people for recruitment to the programme, informing them about the procedures prior to testing, as well as about the result of the test afterwards, counselling about

---

[42] For instance mammography programs (see Section 5.3).

possible follow up-procedures, and help with coping with the reactions to the test result. Even if defensible in terms of the disease targeted, the testing method utilised, and the treatments available, a programme may still be open to serious criticism if organised in an inferior way as regards these processes.

There is a growing consensus that also screening has to apply procedures of informed consent, i.e. disclosing information about the testing and its possible consequences in a way relevant and understandable to the participant and without applying autonomy-restricting pressure to participate, at least if the programmes target adults or, in other cases, if not benefits very clearly outweighs drawbacks. This implies that it is problematic to screen for a large number of disorders simultaneously, due to the risk of information overload. It also implies that screening programmes need to consider questions regarding secondary or unexpected information, as well as the potential conflict between promoting public health and respecting individual autonomy.

In order for individuals to benefit from the medical information resulting from a screening programme, what information is disclosed and how it is disclosed is of great importance. Counselling is a practice founded on the norm of non-directiveness that aims at designing the situation of disclosure so that it is conducive to the health, well-being, and autonomy of the individual. It is not an easy affair to achieve these goals; it takes time, effort, training, and, thus, resources. This presents a special problem for screening programmes, partly because they involve such large quantities of people, partly because these people are as a rule not very prepared for the prospect of testing themselves. The problem of benefiting and not harming the patient to a satisfying degree to an acceptable cost that this creates for screening programmes at least places the onus probandi on the one proposing such a programme: benefits must clearly outweigh harms in order for the implementation of a screening programme to be justified, and securing that so is the case must not make the programme too costly.

This gives rise to difficult questions regarding the cost-effectiveness of screening programmes, as well as questions of how to balance efficient uptake of participants with respect for their autonomy.

Chapter 5
Case Studies

So far, we have been considering numerous factors that are relevant to answer the main question of this book: what screening programmes are morally justified? As we have seen, this hinges on so many factors that it may be difficult to see the full picture. In this chapter, we would like to bring these factors together in order to assess different screening programmes that are implemented or suggested. We will do this by considering four contested (types of) programmes. Some of these are already up and running, while others are in development or, at least, seriously contemplated for the future.

The number of screening programmes up and running in the world are plenty. They become even more numerous if we include suggested ones. A selection has to be made in order to make the discussion manageable. But why have we selected these ones? Primarily because they are of current interest, controversial, and telling regarding the general debate on which screening programmes are defensible. Non-invasive prenatal diagnosis (Section 5.1) is the next frontier of prenatal diagnosis. Screening for fragile X (Section 5.2) is, similarly, an instance of the next frontier for neonatal screening, since it is an instance of screening for a disorder for which there are few medical benefits of screening. In this regard, it is a case that illuminates the wider question of which benefits that are relevant for assessing (neonatal) screening programmes.[1] Mammography screening (Section 5.3) and PSA screening (Section 5.4) are perhaps the most widely known

---

[1] The careful reader will note that we have represented cases of prenatal, neonatal, and adult screening, but not adolescent screening. The reason is that adolescent screening programmes are rarely suggested (although there are exceptions, e.g. ATD-screening, see Section 3.3.2) and even more rarely implemented. One notable exception is HPV (human papillomavirus). However, then it is seldom screening in isolation that is discussed but screening in conjunction with vaccination, which brings a host of ethical problems of their own (Dawson, 2007). For an excellent discussion of these problems, see Malmqvist et al. (2010).

and debated examples of screening programmes.[2] They are also useful to illustrate the social logic and pragmatics of screening institutions.

5.1 Non-invasive Prenatal Diagnosis

Non-invasive prenatal diagnosis is a testing method in its infancy. Hence, we will start this section with a fairly detailed presentation of it. Simply put, non-invasive prenatal diagnosis (NIPD) is genetic testing of the foetus performed through analysis of components in the pregnant woman's blood. This means that no invasive measures into the uterus (such as amniocentesis or chorionic villus sampling (CVS)) are needed for the test and, hence, that the risk for miscarriage (0.5–1%) associated with such methods is avoided. NIPD shares this feature with the new prenatal risk assessment instruments mentioned in Sections 3.1.1 and 3.2.1 and also with routine obstetric ultrasound. However, unlike those methods, NIPD aims at delivering very precise and specific information about the presence or non-presence of disease or genes predisposing for disease. More specifically, NIPD currently usually refers to testing of the genetic status of foetuses by analysing cell-free nucleic acids (cffNA) in the maternal circulation during pregnancy. At present, one cannot isolate foetal DNA or RNA that is identical to maternal cffNA (which regards 90–95% of the cffNA in the pregnant woman's blood). Thus, primarily paternally inherited foetal DNA and unique DNA from the placenta can be analyzed by NIPD in its present stage of development. This, in turn, implies that contemporary NIPD has only limited applications. In the clinical setting, NIPD is only used at some specialist centres for sex determination and RHD typing, and foremost the latter.[3]

However, already today, paternally inherited dominant disorders, e.g. some cases of Huntington's disease, as well as recessive disorders with a difference between paternal and maternal mutation, e.g. some variants of cystic fibrosis, can be detected by NIPD. Moreover, experts in the field assess it as very likely that most chromosomal disorders, including Down syndrome, will be identifiable by NIPD within the next few years.[4] In fact, there is evidence already today that NIPD for Down syndrome is clinically efficacious

---

[2] To go on adding similar cases of adult screening programmes, e.g. colon cancer screening, would only lead to tiresome repetition. We trust that interested readers would be able to extrapolate from the cases presented.
[3] Wright and Burton (2008). RHD typing can be used to avoid maternal immune response in RHD-negative women carrying an RHD-positive foetus.
[4] See e.g. Wright and Chitty (2009).

and practically feasible.[5] In the longer run, all monogenetic disorders will possibly be possible to detect through NIPD as well, given that the PCR technology is developed further.[6] This paves way for a radical shift in the prerequisites of prenatal diagnosis. Not only can the medical risk of prenatal diagnosis be, in practice, eliminated. NIPD seems to be as accurate and reliable as the current invasive methods, with virtually no false positives or negatives.[7] In other words, most future prenatal testing can be as precise and predictively powerful as invasive prenatal is today, while being as safe as the pricking of a finger. These advantages with NIPD are obvious.

Accordingly, it should come as no surprise that NIPD prenatal *screening* has already been suggested.[8] So strong is the pull of this notion that it has been claimed that NIPD should be introduced "even if the demand for it is initially so high as to prevent comprehensive pre-test counselling."[9] In light of the obvious advantages with NIPD as compared to invasive prenatal diagnosis, it may seem obviously unethical not to use NIPD whenever possible. That is, even if the understanding of patients of what is thereby embarked upon has not been reasonably secured.

However, before getting ahead of ourselves and give in to this, admittedly, persuasive argumentation from the advocates of NIPD, we need to apply some of the basic distinctions developed in earlier chapters. First, there are questions regarding moral problems of prenatal diagnosis in general. Those are well-documented and discussed elsewhere and while the medical safety of the testing methods is surely one of these, it is far from the only one (see Section 3.1.1). In this context, we are facing arguments that arise from considering NIPD particularly, *as compared with other prenatal diagnosis*. This means, for instance, that a possible positive conclusion regarding the use of NIPD does not mean that NIPD should be used; only that its use is preferable to some other methods for prenatal diagnosis, which in turn may be the subject of much criticism that remains even if the risk for miscarriage due to testing is eliminated.

Second, we are focusing on the use of NIPD for *screening* purposes. What we are asking is, then, not if NIPD should be used in prenatal testing practices in general, but if such use – given that it is a defensible method for prenatal testing at all – *should be organised as screening*. This question,

---

[5] Chiu et al. (2011).
[6] Lun et al. (2008).
[7] For instance, the false negative rate in RHD typing by NIPD is estimated to 0.2% today, and can be expected to be reduced further (van der Schoot et al., 2008).
[8] Wright and Burton (2008), p. 147.
[9] Ravitsky (2009), p. 516.

in turn, can be approached in two ways: (1) By asking whether or not the criticism against prenatal screening described in Sections 3.1.1 and 3.2.1 is somehow avoided by NIPD, so that the rather heavy reasons against organising prenatal testing as screening disappear. (2) By asking whether or not the use of NIPD would *be preferable to* the current methods applied for prenatal screening in light of the just mentioned criticism. The relation between (1) and (2) is that while a positive answer to (2) is a *necessary* condition for having a positive answer to (1), it is *not* sufficient. That is, even if NIPD has clear advantages compared to currently applied screening methods, the use of NIPD for screening may still be unjustified.

Third, one has to ask: screening for *what*? In order to make the discussion manageable and practically relevant, we are focusing on *certain kinds* of prenatal screening. No one has suggested that NIPD should be implemented as a screening for everything it *could* be used for (which would likely be, in principle, all detectable chromosomal and monogenetic diseases within a foreseeable future).

For these reasons, we will, in line with most ethical debate in the area, confine ourselves to the most likely application of NIPD in a screening setting (at least in the short run), namely as a substitution for non-invasive risk assessment methods currently organised as screening in many countries (described in Sections 3.1.1 and 3.2.1) with invasive prenatal diagnosis as a possible follow-up.[10] That is, we are discussing the procedure for detecting primarily Down syndrome implemented in many developed countries, e.g. Denmark, Sweden and Germany, where maternal serum testing is combined with ultrasound and maternal age in order to stratify a high-risk group that, over a predetermined cut-off point, is offered to continue with invasive diagnosis – henceforth called *standard prenatal screening*. The first question is, then, if NIPD would be preferable to standard prenatal screening, assuming that it can detect all of the conditions for which the risk can be assessed by current methods. The second question is: If so, is NIPD preferable *enough* to actually warrant screening in light of the criticism against standard prenatal screening discussed earlier.

We have seen that the obvious benefit with NIPD is its safety – it avoids the risk of miscarriage due to invasive methods. This, however, is an advantage that is *shared with* standard prenatal screening. At best, then, this particular feature may be an argument for introducing NIPD as the follow-up procedure instead of invasive prenatal diagnosis. What other advantages of NIPD are there?

---

[10] de Jong et al. (2010), p. 273.

5.1 Non-invasive Prenatal Diagnosis

We argue, in several places in this book (see e.g. Sections 2.2 and 3.1.1) that autonomy is unlikely to be a good justification for screening in general (at least not in conjunction with requirements of cost-effectiveness). Despite this, as we have seen, reproductive autonomy is sometimes invoked as an argument, not only for prenatal testing, but also for prenatal screening. As we argue in earlier (Section 3.1.1), this is not a very plausible argument for prenatal screening today. However, if there are some distinguished advantages from the point of view of autonomy with NIPD, it might be held that autonomy is promoted by NIPD screening or, at least, that the autonomy-related problems with screening are reduced to a considerable extent by using NIPD. So, are there any autonomy-related benefits with NIPD?

There are at least two autonomy-related possible benefits of NIPD.[11] First, NIPD may extend the period of time for decision making. It is possible to extract cffNA already at the 4th week of pregnancy, although at present, reliable testing can be performed at the 7th week.[12] Even then, however, it is at least 3 weeks earlier in pregnancy than standard prenatal screening. This fact has been used to argue that NIPD "would therefore enhance reproductive autonomy by allowing women more time to gather information."[13]

This may be true to some extent, but nevertheless it is not a very strong argument. It is important to distinguish between different decisions. NIPD is unlikely to extend the period of decision making as regard the decision of whether or not to have prenatal testing at all very much. On the contrary, the possibility of earlier testing may rush the decision to have testing, which, in effect, would make the period for deciding whether or not test shorter. The time to ponder that decision for the individual couple starts when they are made aware of their alternatives for prenatal diagnosis and ends when the possibility to have prenatal diagnosis has run out or become practically meaningless (e.g. for further decision making regarding termination of pregnancy). If there is a screening programme for all pregnant women, couples should be counselled and asked to participate very soon after the fact of pregnancy has been established regardless of if the method utilised in the programme is NIPD or standard prenatal screening. Thus, the fact that NIPD can be performed earlier may shrink the timeframe for the decision on participation and, as a consequence, makes it more difficult to provide proper counselling.

---

[11] Ravitsky (2009).
[12] Wright and Burton (2008).
[13] Ravitsky (2009), p. 733.

However, due to the potential for earlier diagnosis, NIPD could extend the period between having prenatal diagnosis and deciding what to do as a result of it. Most countries have a preset limit for legal abortion and, where not, it becomes increasingly more difficult to obtain a legal abortion as pregnancy proceeds. So, for instance, the time available for deciding whether or not to terminate pregnancy could be prolonged by NIPD. This, at least, harbours the possibility of more informed and carefully thought through decisions on this particular topic, which can be beneficial from the point of view of autonomy.[14] Moreover, there are potential benefits in terms of well-being with earlier diagnosis since, generally, abortion is likely to be less traumatic the earlier in the pregnancy it is performed (if that is what one decides as a result of prenatal diagnosis).

A second autonomy-related possible benefit with NIPD is that it eliminates one important factor in the decision of whether or not to perform prenatal testing, namely the risk of miscarriage. This means that prospective parents can concentrate on the relative pros- and cons for them to have prenatal diagnosis without factoring in that risk.[15]

This connects to a third benefit from the point of view of autonomy. As remarked earlier (Sections 3.1.1, 3.2.1 and Chapter 4) standard prenatal screening has a significant downside in that the information provided is rather complex (built, as it is, out of accumulating risk figures from a number of sources into a single one). In addition, there are problems relating to false positives and negatives to consider. As mentioned, specialists on standard prenatal screening have held out these features as a challenge for standard prenatal screening to be able to offer counselling procedures of sufficient potency to make the screening ethically defensible.[16] On both these counts, however, NIPD would seem to be much easier to handle. First, the indications are that this method would suffer almost no problems with false results of any kind. Second, once NIPD is validated, the results delivered will be of the same straightforward binary nature as when using current invasive prenatal testing techniques. True, there will still be many important things to address in counselling, but two major complications would seem to be eliminated.

This far, it looks like NIPD is clearly better than standard prenatal screening. However, there are also potential autonomy-related drawbacks with NIPD. The mentioned facts that NIPD allows for earlier diagnosis and

---

[14] Juth (2005), chapter 3.
[15] Ravitsky (2009), p. 733.
[16] Saltvedt (2005).

eliminates the risk for miscarriage could both be held to be problematic, since both these facts likely lowers the psychological threshold to use both prenatal diagnosis and abortion (as a result of a positive test result).[17] The point here is not that this is morally problematic, since it increases likelihood of more prenatal diagnosis and subsequent abortion. There are countless opinions on the extent to which early abortions in general and so-called selective abortions in particular are moral problems at all. The point is rather one about autonomy: reducing psychological barriers may make these decisions be made more frivolously and less thought through. This is a known phenomenon, e.g., in genetic breast cancer testing, where counsellors may need to hold back patients who want to proceed directly to a blood test and make them reflect on the possible long-term consequences of testing.[18] For that reason, counselling may become more challenging and have to be strengthened in other aspects, thus possibly counteracting the counselling-related benefits mentioned in earlier paragraphs.

Moreover, standard prenatal screening is a two-step procedure (first risk assessment, then invasive testing), while NIPD is a one-step test. This means that standard prenatal screening offers a respite for decision making and, more importantly, that couples meet the physician or other counsellor at least twice.[19] That is, standard prenatal screening offers more natural room for consultation and information exchange than NIPD.

Of course, it is possible to design NIPD so that it contains as ambitious pre-test counselling as standard prenatal screening or, preferably, more ambitious.[20] For instance, pre-test prenatal counselling could adopt the model from adult counselling, with a separate meeting before decision making. This could perhaps counterbalance the mentioned autonomy-related drawbacks, at least to some extent.

However, there is one more autonomy-related problem that seems even more difficult to tackle, namely that NIPD may be considered an "offer you cannot refuse".[21] The fact that invasive prenatal testing carries a risk for miscarriage provides women who are reluctant to perform prenatal testing a straightforward rationale to abstain. When this risk is eliminated, they may feel pressured to accept the offer to use prenatal testing. This is not only

---

[17] Schmitz et al. (2009).
[18] Juth (2005), p. 105.
[19] Schmitz et al. (2009).
[20] There is plenty empirical research to demonstrate that procedures of informed consent are less than satisfactory when it comes to standard prenatal screening, see e.g. Renner (2006); and Favre et al. (2007).
[21] Schmitz et al. (2009).

a farfetched speculation; an American study showed that almost a third of responding women (28.5%) said that they would feel compelled to accept an offer for NIPD.[22] This is perhaps not so surprising after all. In the light of a growing tendency to see one's health as one's own responsibility,[23] it might be considered irresponsible to voluntarily forego the option to find out whether one's children can be seriously ill or disabled, especially if this can be done in a safe and reliable way. Add to this the various societal or other structural pressures that are not seldom present in the case of prenatal screening, and the fact that that NIPD is easier to choose may function to strengthen the sort of criticism against prenatal screening that has been launched by the disability movement (see Section 3.1.1).

In sum, it is presently hard to take a stand on whether or not NIPD would be a benefit from the point of view of autonomy. Especially if organised as screening, there is reason to suspect that prenatal diagnosis could be considered the "default position", i.e., that testing is taken for granted and considered "routine". Such a process is sometimes called the "normalisation"[24] and is considered a risk, since it increases pressure to have testing. Such a development would counteract rather than promote autonomy. The concern that such a development-PCR would actually take place is strengthened by a study amongst health care professionals in the UK. In this study, a significant proportion of respondents stated that less strict procedures for informed consent are needed for NIPD than for invasive testing.[25] The tendency towards accepting more of pressure on people, then, is not only of the subtle and structural kind. Health professionals in fact seem prepared to endorse it!

Moreover, NIPD screening may accentuate other general problems with prenatal diagnosis and screening. One such example is the problem of specification creep,[26] i.e., the tendency in many health care systems that care is demanded and supplied for increasingly less serious conditions, or even problems that are presently not considered as medical conditions at all. We noted this development in prenatal diagnosis (in Section 3.1.1) due to the introduction of the QF-PCR method, which is designed to target several comparably mild conditions while leaving out some very severe ones. NIPD could give rise to similar scenarios of starting to slide along a slippery

---

[22] Zamerowski et al. (2001).

[23] At least this is clear in the bioethical literature, e.g. from the growing controversies about luck egalitarianism, see Segall (2010).

[24] de Jong et al. (2010), pp. 273–274.

[25] van den Heuvel et al. (2009).

[26] Hall et al. (2010), p. 250.

5.1 Non-invasive Prenatal Diagnosis

slope of pathologies. For instance, NIPD can be used for sex determination in order to diagnose serious sex linked conditions, such as haemophilia. However, it can also be used to diagnose less serious conditions, such as triple X syndrome. Moreover, it can be used only to select for preferred sex without any connection to health considerations.

NIPD has actually been used for diagnosing foetuses with haemophilia. Even this can be considered as an instance of specification creep, since haemophilia is seldom investigated by invasive prenatal diagnosis.[27] This is the point: due to NIPD's accuracy, safety, and potential for early diagnosis, conditions which never would be considered as serious enough for invasive methods suddenly can be considered easy and self-evident targets for prenatal diagnosis. The problem with such developments is complex. One part of it has to do with an invisible, hard-to-detect shift taking place in the basic values driving the practice of prenatal diagnosis; moving from a health care ethical perspective to a more public health oriented one. For sure, strict adherence to ideals of respecting autonomy may soften the consequences for prospective parents of such a shift.[28] However, we have also seen (in Chapter 2) that this ideal is considerably weakened when placed in a public health context.[29] Another part of the problem is about having mere technological development dictate how large portion of health care resources are allocated to prenatal screening. From society's point of view, this is an obvious downside in terms of both economics and ethics.

Another problem that NIPD screening may accentuate regards stigmatisation and discrimination of people with the conditions screened for (these problems are discussed at length in Section 3.1.1). In this section, we focus on NIPD screening replacing standard prenatal screening. However, standard prenatal screening targets certain chromosomal aberrations, and especially Down syndrome. Because of the safety and accuracy of NIPD, the prevalence of Down syndrome is likely to be further reduced if it replaces standard prenatal screening.[30] This may strengthen the impression that prenatal screening is primarily a vehicle to avoid individuals with certain conditions, rather than a tool to promote reproductive autonomy.

The only way to avoid such "directive" prenatal diagnosis seems to be to offer NIPD for a broader range of conditions, without any predetermined lists or criteria on what to look for (see Section 3.1.1). However, this seems

---

[27] Hall et al. (2010), p. 250.
[28] Munthe (1996), chapter 4.
[29] Munthe (2008).
[30] Benn and Chapman (2010).

practically incompatible with using NIPD as *screening*, at least if one would like to maintain acceptable standards of informed consent.[31] It is hardly conceivable to inform about everything that could be detected by NIPD to all pregnant couples, at least if the goal is not only to disclose information but rather to have the patient to *understand* it. It should be bourn in mind that this is a known problem in prenatal screening already today.

This means that NIPD screening faces a dilemma. Either it is similar to the standard prenatal screening of today and becomes directive, with all the related problems of autonomy. Or it becomes practically inconceivable or, at least, very hard to square with plausible requirements of cost-effectiveness.

Having said this, we do not want to deny that (a developed and duly validated version of) NIPD would be superior to invasive prenatal diagnosis in several ways, most obviously because of its safety. However, this rather indicates that the idea of using NIPD for prenatal *screening* is a mistake to start with. Just as with the new non-invasive risk assessment methods, NIPD is a very welcome addition to prenatal *testing* (undertaken at the initiative and request of individual prospective parents) – at least if it is introduced with awareness of its potential problems.

5.2 Neonatal Screening for Fragile X

Fragile X is the most common hereditary cause of mental disability.[32] Despite that, there are no screening programmes for fragile X anywhere in the world. Since Down syndrome is a target for so much testing and screening, although primarily prenatally, one would perhaps expect fragile X to be so too. In fact, prevalence, taken in isolation, makes a stronger case for screening for fragile X than any of the traditional metabolic disorders targeted by neonatal screening today. The fact that fragile X is not screened for is likely due to the lack of an easily applicable and sufficiently reliable test for this syndrome. However, experts believe that such a test will be available, and when this happens "advocacy groups will inevitably pressure states to include these conditions in state NBS [new born screening] programs."[33] It is not unlikely that parts of the medical community will provide pressure as well, considering the inherent "institutional logic" of screening (see Chapter 1 and the next section), e.g. the various forms of pay-off that come

---

[31] See Section 5.1 and Deans and Newson (2010).
[32] Connor and Ferguson-Smith (1997), p. 137.
[33] Bailey et al. (2008), p. 699.

5.2 Neonatal Screening for Fragile X

with administering such programmes. The question, then, is: should society yield to such pressures or, more neutrally put, ought neonatal screening for fragile X to be implemented when there is a test of sufficient quality?

Fragile X syndrome is a monogenetic X-linked disorder that occurs due to a mutation which increases the number of CGG triplets in the FMR-1 gene. Normally, there are 5–50 repeats of this triplet, in the case of a so-called premutation there are 50–200 repeats, and, in the full mutation, 200 repeats or more. The likelihood of premutations expanding to full mutations increases with successive generations. The full mutation can lead to a wide variety of symptoms, foremost different degrees of intellectual disability, but also connective tissue problems, attention deficit disorders, autistic behaviour and compulsive tics, such as hand-biting. The penetrance of the syndrome is usually much milder for females, since it is X-linked; the other X-chromosome compensate for the mutated one. Male carriers of the premutation, which as a full mutation gives rise to fragile X, can develop FXTAS, a deteriorating neurological disorder, with onset in late middle-age or old age and symptoms similar to Parkinson's. While females are often asymptomatic carriers, their children have a 50% likelihood of inheriting the mutation. Sons can never get the mutation from their fathers, but daughters always do, in which case they most likely become healthy carriers.

So, ought neonatal screening for fragile X to be implemented when there is a reliable test? A way to start to tackle this question is by comparing neonatal screening for fragile X with neonatal screening programmes that are almost unanimously held to be warranted, i.e. PKU-screening as it is performed in most developed countries. The most notable difference between fragile X and PKU (and similar metabolic disorders, see Section 3.1.2) is that there is no preventive or curative medical intervention for the former. Since such interventions have been considered a necessary prerequisite for considering the implementation of neonatal screening in the first place, one can ask why screening for fragile X should be suggested at all. Nevertheless, it has actually been suggested.[34] Moreover, suggestions in the same vein may be expected to increase as the quality of tests improves. So, what *reasons* are there to take such a suggestion seriously?

The most obvious argument in favour of fragile X neonatal screening is that early detection may have some benefits. These are partly therapeutic, but mainly about psychosocial advantages for the family and, to some extent, general social benefits. The therapeutic benefits of having an early

---

[34] See e.g. Skinner et al. (2003).

diagnosis for fragile X are undisputed, since it allows for parents to adjust their behaviour towards the child in light of a correct view of his (or, less often, her) problems and to provide appropriate training as soon as possible.[35] However, it is uncertain to what extent screening will actually result in earlier detection. Since there have been no screening programmes for fragile X so far, there is simply no empirical data regarding this yet. As far as psychosocial benefits goes, early detection could prevent the sort of "diagnostic odysseys", where parents have to go from doctor to doctor for a long time in order to have a specific answer regarding what is wrong with their child. Besides burdening the parents, this also delays the application of appropriate adaption to the child's special needs, as well as adds unnecessary costs to society. In addition, besides preventing problems, earlier diagnosis can contribute to research through a better understanding of the earliest phases of this condition.

These alleged benefits lead to two questions: to what extent should they be considered relevant at all for introducing neonatal screening (in the absence of effective prevention or cure)? Do these benefits outweigh the downsides?

Let us start with the former question. Here, we run into a discussion analogous to the one regarding treatment: some claim that counselling should be seen as treatment (Chapters 2 and 4), primarily in prenatal diagnosis (Section 3.1.1) due to its potential for promoting well-informed reproductive decisions. Similarly, it has been claimed that the concept of benefit in neonatal screening should include the mentioned psychosocial and social benefits, as well as *future* reproductive autonomy.[36] As pointed out early on in this book, however, one must not be tricked by the impression of this question as being merely semantic. The issue is not simply what words to apply to the imagined follow-up procedure in a screening programme for fragile X, but a *moral* one: the question is if the suggested considerations are sufficient to warrant that a neonatal screening programme for fragile X would be able to provide sufficiently beneficial follow-up procedures.

---

[35] While, e.g., the immediately visible autism-type symptoms of fragile X may be noticed rather soon, it is of both psychosocial and therapeutic importance to know whether the symptoms are caused by fragile X or not. For instance, while the classic sign of lack of eye contact, physical touch and verbal interaction may, in the general autism case, often be due to a straightforward inability of the child (possible to gradually repair through special training and adapted surroundings), in the specific case of fragile X, it rather seems to be due to these basic social and communicative behaviours causing discomfort, therefore being rationally avoided by the child. See, e.g., Dew-Hughes (2003); and Jenssen Hagerman and Hagerman (2002).

[36] Already in ACMG's framework, these considerations are included (Baily and Murray, 2008, p. 28). See also Bailey et al. (2005a).

5.2 Neonatal Screening for Fragile X

One problem with making a generous interpretation of benefits and answer this question affirmatively is that it makes it difficult to see what *not* to screen for. In principle, any trait can have psychosocial consequences and any hereditary trait could be considered relevant for future reproductive decision making, e.g. depending on the preferences of the parents. Even if one manages to defend the idea that screening is warranted only for diseases, and not just any trait,[37] already today over 8000 monogenetic diseases have been identified. If we add chromosome aberrations and multifactorial diseases, the sheer number of potential conditions to screen for is hard to grasp. However, if we invite any sorts of potential benefits as a justification of screening, it is not easy to say where to stop: "if we screen for fragile X, since it reduces diagnostic odysseys and helps future family planning, why not also screen for cystic fibrosis? And if for cystic fibrosis, why not for Lesch-Nyhan, Marfan, etc.?" This is a reason for applying rather strict requirements on what sort of follow-up procedure and benefit that may warrant screening. This, by itself, however, does not settle the fragile X case.

For the sake of the argument, therefore, let us for now assume that a convincing case in terms of follow-up procedures and benefits can be made for including fragile X in neonatal screening, while most other genetic disorders are kept out. That is, the relative benefits of screening for fragile X is greater than for most other genetic diseases for which there are no effective medical measures. This still leaves us with the other question: do the benefits of neonatal screening for fragile X outweigh the downsides?

Routine neonatal screening, e.g. for PKU, have no or very loose procedures for informed consent – the standard is that neonatal screening is presented as just another part of routine neonatal care.[38] However, when there is no effective medical intervention and the benefits are mainly psychosocial and autonomy-related, it is especially important to obtain informed consent (Sections 2.2 and 4.1). Because of this, also those inclined to support neonatal screening for fragile X emphasise the need for proper consent procedures.[39] This would include e.g. information about the screening procedure, potential risks and benefits of diagnosis, explanation of fragile X (how it is inherited, risk of recurrence, diagnosis and prognosis, interventions), and how to share information with relatives. Moreover, one has to ensure that the information is not only available or presented, but also

---

[37] It has proven very difficult to draw a morally relevant distinction between disease and other properties (Juengst, 2003).
[38] Baily and Murray (2008).
[39] Bailey et al. (2008).

properly understood by the parent(s). As argued in Section 4.1, this implies the application of rather ambitious counselling procedures.[40] The specifics of the situation in which neonatal screening is carried out, however, suggests a need for even more demanding counselling efforts.

In order to be effectively executed, neonatal screening normally is performed during the post delivery hospital stay, where, for explanations easily understood, parents have only vague conceptions about information they receive, which increases difficulties of understanding.[41] An alternative would be to provide information during pregnancy, although it is difficult to see how that could be done without raising anxiety for the health of the coming baby. Moreover, the risk for information overload seems obvious when considering all the information about pregnancy itself and the imminent birth that needs to be digested by the parents. Accordingly, it is difficult to see how neonatal screening aimed at targeting the whole population of parents (as with PKU) could achieve acceptable standards of informed consent, and even harder to see that it could do so in a cost-effective manner. Counselling by appropriately trained personnel that are given appropriate time at their disposal is costly indeed. As mentioned elsewhere, this is an ethical concern, since opportunities to benefit are foregone if resources might be put to better use elsewhere.

However, the consent and counselling related problems with screening for fragile X is not "only" a matter of (economic) cost and the risks of confusing and manipulating people. Another problem is the phenomenon of secondary or incidental findings, i.e., identified conditions that are not intentionally sought for. Some incidental findings are unavoidable if a certain test is done. For instance, the tandem mass spectrometry proposed by ACMG, identifying 29 conditions (see Section 3.1.2), necessarily identifies 25 secondary conditions that do not meet the criteria for screening. Similarly, a fragile X test would be very likely to identify other X-linked disorders, which are not nearly as serious as regards symptoms and medical problems as fragile X, e.g. Turner, Klinefelter's, XXY, and triple X syndrome.[42] Especially the two latter rarely present any symptoms at all, and most carriers live their lives happily ignorant about the "aberration". Population screening would likely result in an increased identification of these disorders, with no medical advantages, but potential problems with

---

[40] Bailey et al. (2008), p. 697.
[41] Davis et al. (2006).
[42] Personal information from geneticist Erik Björck, Karolinska institute. See also Bailey et al. (2008).

disturbed family relations, negative self-concept, societal stigmatisation, and insurance or employment discrimination.[43]

Moreover, as regards fragile X itself, many people carry *pre*mutations that may cause medical problems later in life or to subsequent generations, or are asymptomatic female carriers. For these, there are no therapeutic advantages and no diagnostic odysseys to be avoided, but only the risk of all the mentioned drawbacks. Possibly, refined testing methods may be developed that are able to exclude premutations. However, it is difficult to envision such a development without the cost of an elevated probability of missing full mutations. That is, also here we encounter the problem of trading off the false positives and negatives of a test.

Yet another psychosocial problem is how to handle the informing of relatives. The metabolic disorders usually screened for are autosomal recessive traits, which makes it highly unlikely that the relatives of the tested child will have children with the disease in question. However, since fragile X is an X-linked disorder, it has a stronger pattern of heredity. Accordingly, it is of relevance for both close and extended relatives to find out whether they are carriers.

The problems with informing relatives are well known in clinical genetics and genetic counselling.[44] For instance, should relatives that are potential carriers be informed that a test has been performed and/or the results of the test? In what circumstances? Who should provide this information? The tested person or some health care professional? Should health care professionals try to persuade the tested person (or, in the neonatal case, the parents) to disclose the information if the this person is reluctant to do so? To what extent and in what circumstances (if any)? If the he or she still refuses to disclose, are there reasons for health care to inform the relatives anyway? In what circumstances (if any)? Do relatives have a right *not* to receive the information that a test has been performed and/or the results of the test? If so, how does the programme need to be organised so that this right can be effectively respected?

These questions are not unique to neonatal screening for fragile X, but accrue most genetic testing. Hence, one study revealed that a majority of parents who informed relatives experienced difficulties in disclosing information about the disease to relatives, including denial, blame, and anger from the relatives.[45] Of course, if it is determined that relatives should be

---

[43] Bailey et al. (2008), pp. 697–698.
[44] Juth (2005), chapter VI.
[45] Bailey et al. (2005b).

informed at all, which is far from self-evident,[46] health care professionals could take on the responsibility of informing them. The problems are thus accentuated due to the scale of screening programmes. In the clinical setting, they can often be handled on a case by case basis through a dialogue between counsellor and counselee. Having such counselling for the entire population of prospective parents would be a massive and costly undertaking.

These arguments with regard to autonomy-related reasons can be added to the doubts about the follow-up procedure expressed earlier. In all, therefore, the prospect of justifying neonatal screening for fragile X, and consequently any relevantly similar condition, looks bleak indeed.

5.3 Mammography Screening

One hotly contested adult screening programme is mammography for breast cancer. Many developed countries have introduced mammography screening since the 1980s, often targeting women between the age of 45 and 64, who are offered X-ray examination every second or third year (although the age limits vary as does the regularity of check-ups).[47] The explanation as to why mammography screening is contested is a steadily growing criticism from parts of the scientific community.[48] The most influential contemporary critics are representatives from *The Cochrane Collaboration*, an international organisation for evidence based health care, mainly engaged in meta-studies. However, this criticism has not stood unanswered. Most notably, some of those who have done the studies to which the meta-studies refer have reacted strongly.[49]

The controversy between these groups concerns the relevance and magnitude of various harms and benefits suggested to be the upshot of mammography screening programmes in various studies. By the *relevance* of

---

[46] Juth (2005), pp. 362–368.
[47] Gøtzsche and Nielsen (2009), p. 4.
[48] A pioneer in this regard has been Czech, Dublin-based expert in oncology and preventive medicine, Petr Skrabanek, who published a long series of critical appraisals of mammography screening in *The Lancet* and other leading medical journals, as well as instigating general discussions of the ethics of medical prevention, starting as early as 1985 (Skrabanek, 1985). See also Skrabanek (2000).
[49] For instance, Stephen Duffy from the UK, Robert Smith from USA and László Tabár and Lennarth Nyström from Sweden. The following articles together give a fairly good overview of the debate: Duffy et al. (2002) (including discussion); Freedman et al. (2004); Törnberg and Nyström (2009a, b); Gøtzsche and Jörgensen (2009a, b); and Gøtzsche and Nielsen (2009).

5.3 Mammography Screening

these factors, we mean how *important* one considers various harms (e.g. harmful side-effects of the testing method) to be when compared to benefits, e.g. reduced mortality in the population. Issues regarding relevance, therefore, seem to be clearly ethical, since they are about the relative importance or (dis)value of different factors given any extent to which they occur. By *magnitude*, in contrast, we mean facts about the extent to which different harms and benefits really are the results of mammography screening, and such issues appear clearly scientific. Seemingly then, there is a clear-cut distinction between disagreements regarding the ethical foundations for assessing mammography screening and disagreements regarding scientific questions. However, on closer scrutiny, we will see that these theoretically distinct parts are in practice interrelated in the debate as it has actually proceeded.

Several aspects that have already been accounted for in this book are relevant to the issue of the justifiability of mammography screening. One aspect regards the nature of the *testing* procedures (see Section 3.2). Mammography screening programmes fulfil the characteristic of traditional screening programmes in the Wilson and Jungner sense: there is testing in several steps where the initial screening is made to sort out individuals for further testing. The initial step is usually carried out by X-ray testing with poor specificity, i.e. many false positives will be correctly classified only in later stages. Even if eventually cleared from the diagnosis, the psychological effect of receiving a breast cancer diagnosis can be extensive, including severe anxiety and strained relationships.[50]

The extent of false positives is debated (as are most figures in the literature). However, Cochrane Collaboration estimates that one in ten during a 10 year period of mammography screening is a false positive. Moreover, the question of relevance, i.e. to what extent false positives should be seen as a problem, regardless of extent, is also debated. Opinions on this issue are not scientific, but rest on value-judgements about the importance of avoiding false positives. Typically, proponents think that false positives constitute less of a problem than critics: "Is it worse with examination-induced anxiety than premature death in cancer?" is a telling article title from two proponents.[51]

Another aspect relates to *treatment* (see Section 3.3). Some of the tumours of breast cancer never develop into what is called invasive or real cancer, i.e. fast-growing, malign and ultimately lethal cancer. This means that there are women who are identified as *true* positives for breast cancer, but

---

[50] See Gøtzsche and Nielsen (2009), pp. 15–16, for further references.
[51] Törnberg and Nyström (2009b).

who will never have problems with their cancer. Due to the screening programme, some of these women may be diagnosed with breast cancer. This is sometimes called *overdiagnosis*, i.e. diagnosis for something that would have gone undetected, were it not for the screening, and would not have caused any problems. These women are regularly treated with mastectomy, radiotherapy, or chemotherapy (or several of these treatments) since one cannot say which of these will have invasive and which will have non-invasive breast cancer beforehand. When treatment is initiated due to overdiagnosis, it is called *overtreatment*.

One question in the debate regards the magnitude of overtreatment. Figures cited vary from 10% increase of overtreatment due to the initiation of mammography screening programmes,[52] to 30% (which is the figure that critics believe to be a correct estimate).[53] The different estimates are due to a number of factors, e.g. what studies are considered to be of sufficient scientific quality and what data within different studies that are considered to be reliable and unbiased enough. This is the first sign we want to hold out of the trend in this debate of muddling the line between questions of magnitude and relevance. As an external observer, it is difficult not to suspect that one or both of the sides in the debate let their criteria for sufficient reliability and lack of bias be determined by prejudgements regarding the overall quality of mammography screening. If that is the case, what on the surface looks as a scientific controversy about the actual occurrence of overtreatment turns out to contain a hitherto unanalysed complex of value issues. Presumably, these are about what *importance* that should be attached to overtreatment in the first place (regardless of how common or uncommon it is).

The treatment of breast cancer is invasive and arduous. It regularly involves mastectomy (i.e. the removal of breast tissue) to various degrees, radiotherapy and cytostatic chemotherapy. However, early detection of breast cancer can make the treatment less extensive than it had been in later stages. Moreover, early detection increases the likelihood of a successful treatment. Since screening programmes may result in more early detections, some proponents are reluctant to consider overtreatment as a great problem. Thus, in this case, this would be the prejudgement influencing what criteria are set for reliability and lack of bias.

In opposition to those who would see overtreatment as a minor problem, we have statements like: "[r]educing incidence must be the primary goal, with reducing mortality and important but secondary end

---

[52] Zackrisson et al. (2006).
[53] Gøtzsche and Nielsen (2009), p. 2.

5.3 Mammography Screening

point. Mammography unavoidably increases incidence... [which] result in overdiagnosis and overtreatment... Thus real mortality reductions attributable to mammography must be large and secure to justify this possible harm."[54] More radically still, in relation to a discussion of the Wilson and Jungner criteria, one author writes: "note that the WHO principles say nothing about offsetting the harms of screening with benefits. They do not speak of harms at all, evidently because it would be perverse for a test designed for an entire population (principle six) to harm in the name of prevention."[55] That is, *any* harms are difficult to justify within a screening programme, according to this author. Ideas of this sort would then be the prejudgement that influences reliability and bias criteria on this side of the debate.

Our conclusion from all of this is that the debate regarding overdiagnosis and -treatment, while being cast in the language of scientific facts, is powered mainly by deep *ethical* disagreements. Positions with regard to this disagreement range from the idea that overdiagnosis and -treatment is not a great problem at all, to the notion that it is a conclusive argument against a screening programme. These positions can, of course, be upheld almost regardless of the magnitude of overtreatment. However, as pointed out, it is difficult to free oneself from the suspicion that this ethical disagreement has been allowed (by one or both parties) to leak into also what is supposed to be the purely scientific part of the discussion. Such "leakage" is further evidenced by a rather salient rhetorical component of the debate.

The controversy over the relevance of overtreatment has resulted in a dispute regarding the proper *description* of overtreatment.[56] While proponents are inclined to call the negative effects of overtreatment "negative side-effects", critics are inclined to call them "harms". Indeed, critics are inclined to call *overdiagnosis* harm, due to the psycho-social effects of receiving a diagnosis. They are also very clearly prone to cite influential voices in the debate that are inclined to use the term they favour themselves as arguments in favour of their choice of words.[57]

---

[54] McPherson (2010).

[55] Justman (2010). Justman discusses PSA screening for prostate cancer, but the point regarding offsetting harms with benefits is general or at least applies also to mammography, given the great similarities in terms of harms and benefits between the two (see below).

[56] See Törnberg and Nyström (2009b); and Gøtzsche and Jörgensen (2009b).

[57] E.g., Esserman et al. (2009).

However, both sides would do well to remember that this dispute is not scientific or semantic. At the heart of the disagreement is an ethical question: to what extent and in what ways are overdiagnosis and overtreatment problems and what should be done about them? In the debate, the question is turned into a semantic one regarding the proper terminology to use in the description of overdiagnosis and overtreatment. Proponents should acknowledge that both overdiagnosis and overtreatment are problems, however described. Moreover, they should acknowledge that overtreatment defined as treatment for breast cancer that would have gone undetected and would not have caused any problems, were it not for the screening, constitutes significant harm in any intelligible use of the term. Then, of course, the difficult empirical question to what extent mammography screening programmes actually causes overtreatment in this sense remains. Critics, in turn, should refrain from using poor ad hominem arguments in favour of their conclusions: the fact that someone says something, regardless of his or her prestige, is not itself an argument that what she says is correct. This includes opinions about the proper use of words.

On this basis, there is good reason for urging both sides to apply some rigour and critical scrutiny to their choices of assessment and inclusion criteria when judging evidence in the scientific discussion. An elementary point is that this discussion can go nowhere until the debating sides have a clear consensus on assessment and inclusion criteria for evidence. As we will now see, this – as one would have thought – self-evident point is actually worth repeating also in relation to other major parts of the mammography screening debate.

One such part regards to what extent mammography screening programmes are successful in terms of treating women that would have otherwise died, i.e. in terms of reduced mortality. Also here, figures differ radically. Proponents of mammography screening often claim that these programmes reduces mortality by 30%[58] (and sometimes even up till over 50% for some screening programmes[59]), while the critics claim that 15% is a more accurate number.[60] Once again, the different estimates are due to e.g. what studies are considered to be of sufficient scientific quality and what data within different studies that are considered to be reliable and unbiased enough. So, once again, the suspicion of a "leakage" between ethical and scientific issues comes to the surface.

---

[58] Törnberg and Nyström (2009a).
[59] Swedish Organized Service Screening Evaluation Group (2006).
[60] Gøtzsche and Nielsen (2009), p. 2.

5.3 Mammography Screening

In order to take only one example of a point of disagreement; different measurements of success in terms of avoided deaths are considered appropriate: breast cancer mortality, cancer mortality, or all-cause mortality. Those who defend current mammography screening programmes tend to favour breast cancer mortality, while the critics tend to favour the two latter outcomes as the relevant measurement.[61] Although it may seem obvious that reduced mortality in terms of breast cancer is the appropriate measure when determining the success of programmes aimed at this very disease, the critics have claimed that there may be a bias inherent in that measurement. One argument for this is that classification of causes of disease often is a matter of interpretation. For instance, when a patient has several cancers when dying, it is not always self-evident which cancer (if any particular one) should be seen as the cause of death. There may be a propensity to classify those within the mammography screening programmes as dying from another cancer or a wholly non-oncologic cause.[62]

As regards scientific disputes such as this, it is hard for a non-expert to tell which side has the strongest arguments (as it obviously is for the experts themselves). However, an outsider can notice the following about the discourse. Both sides in the debate regularly accuse the other side of being unscientific and biased regarding the interpretation of data, and they both pride themselves with having science on their side.[63] Note that we are not dealing with entirely different bodies of data, but the proper *interpretation* of the same data and the same studies. The critics claim that the proponents interpret data in a manner biased towards screening, due to them being in favour of screening programmes to start with. And similarly the other way around. This at least gives rise to the impression that the stand one takes in the scientific questions may be influenced by ethical opinions regarding the desirability of the programmes, rather than the other way around. At least, the debate would probably gain from bringing implicit value-judgements into the open. As a matter of pure logic, any normative conclusion, e.g. regarding the desirability of screening programmes, has to contain normative premises, e.g. about the seriousness or normative importance of overtreatment.[64] Hence, the matter *cannot* be a wholly scientific dispute. Hence, the best way of stopping or managing the leakage between ethical and scientific

---

[61] Duffy et al. (2002), p. 161; and Gøtzsche and Nielsen (2009), pp. 12–14.
[62] Newschaffer et al. (2000).
[63] See Törnberg and Nyström (2009a, b); and Gøtzsche and Jörgensen (2000a, b).
[64] Salwén (2003).

issues is to unpack and openly debate the basic moral premises implicitly applied in what is presented as a purely scientific discussion. On that basis, it becomes at least theoretically possible to establish the sort of basic methodological consensus essential for any scientific inquiry to get off the ground.

Whatever position one takes in the debate, the way in which it has been conducted further demonstrates the self-reinforcing institutional effects of screening programmes (see Chapter 1). Surveys demonstrate that the general public has great confidence in mammography screening programmes, to the extent that they grossly overestimate their positive effects (whichever figures you believe to be correct). For instance, according to a survey from US, UK, Italy, and Switzerland, 68% of women believe that screening *reduces the risk of being diagnosed with breast cancer* (as we have seen, due to screening-induced overdiagnosis, the opposite is true).[65] Moreover, the same study shows that only 8% of the women are aware that mammography screening may have harmful effects. Such beliefs most likely increase the inclination to accept and welcome an invitation to participate in screening.

The fact that the public overestimates the positive effects of mammography screening may have many, not mutually exclusive, explanations. One may be that the general trust in health care affects the trust in screening programmes. If the public trust health care in general, they are probably inclined to think that screening programmes are initiated for very good reasons and adjust their beliefs about the efficiency and safety of the programmes accordingly.

Another explanation may be that the proponents and representatives of screening programmes excessively tend to present themselves as unequivocal advocates of the best interest of patients – quite in line with institutional theory.[66] One indication that this really is the case is the proponents' tendency to present statistical figures in relative terms when presenting reduced mortality figures and in absolute terms when presenting overdiagnosis or overtreatment figures.[67] For instance, you can read that screening reduces mortality with 30%, but that the risk of overtreatment is 0.1% (if one believes these to be the correct figures), instead of consequently using either absolute or relative figures (saying that it reduces mortality with 30% and increases risk of overtreatment with 30%, or, preferably, that the risk of mortality if

---

[65] Domenighetti et al. (2003).
[66] March and Ohlsen (1989).
[67] Partly to avoid this, the Cochrane Collaboration has developed an alternative information leaflet to women invited to mammography screening, in which all figures are represented in terms of ratios rather than relative percentage. See http://www.cochrane.dk/screening/mammography-leaflet.pdf.

5.3 Mammography Screening

not participating is 0.1% and the risk of overtreatment is 0.1%).[68] For the layperson, this, of course, makes reduced mortality rates look much more impressive than the risk of overtreatment. It most likely also makes the reduction in mortality look bigger than it actually is. As we argue elsewhere, there are good reasons to use absolute figures in the clinical setting (see Section 3.2.3), not least since relative figures say very little unless you are familiar with the figures they are related to, i.e., in comparisons to which the risk is increased or decreased. However, even if one uses relative risk figures, one should at least be consistent. Otherwise, it is difficult to avoid the suspicion that the intention is to "market" screening by playing down the negative effects and emphasising the positive ones.

Furthermore, this, of course, severely complicates the prospective of securing truly informed consent. In light of the documented and widespread misconceptions of the public, and the confusing information provided by researchers, a mammography screening programme that could be defended in terms of autonomy would seem to need a very ambitious pre-counselling organisation.

In spite of all this, the self-reinforcing nature of screening programmes has an important consequence: unless blatantly flawed, they will probably be very difficult to roll back once in place. This has nothing to do with what is medically or ethically well-founded, but follows from the pragmatics of institutions and *realpolitik*. This should be kept in mind when new screening programmes are proposed. Since mammography screening programmes are entrenched, critics have not proposed that they should be rolled back, but rather that the information to women who are invited to them should be more "balanced", telling to larger extent about the risks and de-emphasising the positive effects. In the light of the surveys about the public's belief about screening, it is easy to agree that more information to invited women is needed. However, unless there is more unanimity in the scientific community regarding the magnitude of risks and benefits, we are unlikely to see uniform information to the women. This should also be kept in mind in the future: do not launch new screening programmes unless there is sufficient knowledge about their effects that is based on a solid methodological and ethical consensus.

Having said that, if we were to imagine the situation that mammography screening was proposed to be initiated *today* on the basis of *present* evidence (including the confusing scientific debate), it would not be an easy task to find it to be warranted. First, partly due to the lack of methodological

---

[68] See Törnberg and Nyström (2009a, b); and Gøtzsche and Jörgensen (2009a, b).

consensus, the scientific basis of information is much too uncertain and difficult to interpret. Second, if counselling procedures of an appropriate nature were to be included, this would presumably seriously elevate the cost of the programme.

5.4 PSA Screening for Prostate Cancer

PSA screening for prostate cancer has been ongoing and increasing for two decades in the United States. In general, European countries have been much more cautious with introducing large scale PSA screening.[69]

PSA screening uses PSA (prostate-specific antigen) testing, which measures the level of PSA in the blood. A cut-off point is determined,[70] above which the screened individuals are offered further biopsy testing in order to determine whether the initial findings were due to cancer or something else (e.g. benign prostate enlargement or inflammation in the prostate). PSA screening is therefore screening in the classic Wilson and Jungner sense: testing is performed in several steps, where the initial screening is made to sort out individuals for further testing. In this sense, PSA screening is similar to mammography.

Actually, there are several similarities between mammography and PSA screening. The most common arguments for and against the desirability of the screening programmes are the same: the possibility of reduced mortality due to early detection in favour of screening and overdiagnosis and overtreatment against screening. However, there is considerable difference in *tone* between the debate on mammography and PSA screening. There is a markedly more civil tone in the debate on PSA screening as well as much less quarrel in the scientific community regarding which studies are of sufficient quality and how to interpret the result of these studies. This is partly due to the relative unanimous support for the need of well-designed

---

[69] Again, the are great differences between screening in the US and in Europe seems to be partly due to the differences in health care systems. In the US, screening is driven by the commercial health sector, which, at least partly, likely explains why it was introduced without being officially evaluated first (however, also in the case of the US, PSA investigations amount to screening in the sense defined in Section 1.3). Here different socio-economic structures explain why screening looks like it does, just like in the case of neonatal screening (see Section 3.1.2.3).

[70] Ranging from 2.5 till 4 ng/mL, usually with a lower point the younger the screened individuals are (Neal, 2010).

5.4 PSA Screening for Prostate Cancer

randomised trials in order to determine the benefits and drawbacks of organising PSA testing for prostate cancer as screening programmes. Thus, in the PSA case, it seems that proponents and critics have managed to do the elementary scientific job of establishing terminological and methodological consensus on the basis of shared values.

Only very recently, the results of different such trials are beginning to be published in the scientific literature. In 2008, The Cochrane Collaboration published a review: Screening for prostate cancer.[71] The review found only two trials that were eligible for inclusion to start with. These studies could not establish a statistically significant difference in prostate cancer mortality between screened and control populations. Moreover, they did not assess the outcomes in terms of quality of life and cost effectiveness. However, neither of the studies was found to be of "high quality",[72] which led the authors to conclude that "there is insufficient evidence to either support or refute the routine use of mass, selective or opportunistic screening compared to no screening for reducing prostate cancer mortality."[73]

However, in recent years, this scientific vacuum is starting to be remedied. The most ambitious study is The European Randomized Study of Screening for Prostate Cancer (ERSPC), which was initiated in the early 1990s and includes 182,000 men from seven European countries between the ages of 50 and 74. Results published after a median of 9 years shows a reduced rate of death from prostate cancer of 20%.[74] However, the absolute benefit (0.7/1000 reduction) was small and was associated with a 70% increase in prostate cancer diagnosis. In other words, 1410 men needs to be screened and 48 more cases needs to be treated in order to prevent one death, as compared to the non-screened population. In the light of this, an American review that included the ERSPC study was named with the telling title "Randomized trial results did not resolve controversies surrounding prostate cancer screening."[75]

The most favourable results for screening can be found in a Swedish randomised study, which indicates reduced mortality by 40%, although

---

[71] Ilic et al. (2008). The review did not target PSA testing exclusively, but also programmes using digital rectal examination.

[72] For instance, the Quebec study had very low compliance in the screening group, the Norrköping study reported widely in the media about the study (increasing the likelihood of testing among the control group) and neither study compared outcomes with possible confounders, such as socio-demographic data. Ilic et al. (2008), pp. 11–12.

[73] Ilic et al. (2008), p. 2.

[74] Schröder et al. (2009).

[75] Hoffman (2010).

with remaining high figures of overdiagnosis.[76] However, in comparison to ERSPC, the study is small and results are likely affected by a younger age of screening as well a relatively low cut-off level of PSA, which has led to some doubts about its generalisability.[77]

So what can be, and has been, inferred from all this? First, there is wide agreement on the drawbacks of PSA screening. Much prostate cancer has a slow progression rate and would therefore never present any problem for the patient in question, if gone undetected. Especially in older men, the ratio of overtreatment as compared to reduced mortality is considerable. Moreover, PSA-testing, like mammography, brings a significant risk of false positives. In fact, most men tested positive with PSA are subsequently shown not to have prostate cancer; only 25–35% of men who have a follow-up biopsy are diagnosed with cancer.[78] This relates to the well known negative psychological side-effects of false positives (see Section 3.2.2). In addition, there is the risk of false negatives and, consequently, false reassurance, due to the fact that prostate cancer is often slow growing, which means that PSA levels are normal despite prostate cancer actually being present.

However, new data suggest that the extent of the harms of being diagnosed with cancer have not been sufficiently appreciated in the debate. A large cohort study in USA indicates that a diagnosis of prostate cancer may increase the immediate risk of suicide and cardiovascular death.[79] So even if one holds the most important goal to be to reduce mortality, there may be no clear-cut answer to the efficiency of screening if you take possible side-effects, like suicide or cardiovascular death due to increased stress, into account. So even if some drawbacks are unanimously acknowledged in the PSA screening debate, the debate of the extent and kinds of drawbacks that are relevant for assessing PSA screening is far from settled.

Second, there is a relatively broad agreement that the overall benefits of PSA screening programmes, taking into consideration also drawbacks, is not greater than for mammography screening. According to the most favourable assessments of PSA screening, it is better in terms of reduced mortality, but not in avoiding overdiagnosis, although the extent of overdiagnosis is uncertain.[80] However, most assessments rather say that it is actually the other

---

[76] Hugosson et al. (2010).
[77] Neal (2010).
[78] Smith et al. (1997).
[79] Fang et al. (2010). In fact, during the first 3 months after receiving diagnosis, risk of suicide is almost twice as high as in the control group.
[80] Hugosson et al. (2010).

5.4 PSA Screening for Prostate Cancer

way around: that benefits are uncertain, while the harms are undeniable.[81] In any case, assessments still differ. In the light of this, it is surprising, to say the least, that PSA screening was introduced on a large scale in the United States over 20 years ago, when knowledge about its effects were virtually non-existent.

This relates to the third point: the introduction of PSA testing as a screening procedure in the United States must be seen as a huge *experiment*, given the lack of knowledge of its effects when initiated.[82] This means that standard research ethical guidelines for informed consent should have been applied to research subjects, i.e., those who were subjected to PSA screening. It is generally conceded that patients[83] should always have the opportunity for informed consent when tested and treated, but that this is *especially* important in research, due to the very fact that benefits and harms are more uncertain before the research in question is concluded.[84]

Fourth, there is relatively large agreement that standards of informed consent in PSA screening, at least in the United States, have been poor. A recent survey among test subjects revealed that a majority (71.4%) were told about the upside of PSA testing, while only a minority (32%) were informed about potential drawbacks.[85] This shows that the information-part of the informed consent were deficient. In conjunction with the fact that a majority were recommended to take the PSA test,[86] it is safe to conclude that there is, in general, considerable pressure from the health care system in the United States to opt for PSA testing. The survey reminds about testimonies regarding mammography, where patients are told that "it saves lives" and little or none about its drawbacks. In consequence, there is growing consensus that standards of informed consent in PSA screening should be improved, e.g. by addressing drawbacks as well as potential benefits. In fact, more and more organisations in America no longer recommend, or even advice against, PSA

---

[81] See Justman (2010) for more references on this.

[82] This point is repeatedly made in the literature, but deserves to be mentioned again (Justman, 2010).

[83] With a few exceptions, e.g. the decision incompetent. See Beauchamp and Childress (2001), chapter 3.

[84] World Medical Association (1964–2008).

[85] Hoffman et al. (2010).

[86] 70.4% for those who reported very good or excellent health and 78.1% for those who reported worse than very good health (Hoffman et al., 2010, p. 1616).

screening, at least as a general routine screening for all males above a certain age.[87] In light of the American experience and the growing controversy on mammography, European countries would seem to have good reasons indeed for sustaining their reluctance to introduce PSA screening.

As these points demonstrate, there is larger general consensus in the debate on PSA screening than in the debate on mammography. As noted earlier, the tone of the debate is more civil, at least in the scientific community. One of the most conspicuous differences between mammography and PSA screening is that the former already has been implemented for decades, while the latter to a large extent is confined to randomised studies, at least in Europe. That is, there is to a lesser extent a heart- (or, for that matter, budget-) felt need to defend PSA screening, since it is implemented to a lesser extent to start with. This lends support to the previously presented hypothesis that screening often becomes a self-reinforcing institution, where representatives of screening soon develop into its fierce advocates. Also, it lends support to the hypothesis that value-judgement regarding the desirability of programmes may influence interpretation of scientific data, rather than the other way around. Most importantly, the debates on both mammography and PSA screening demonstrate that science alone cannot settle the issue of to what extent these screening programmes are warranted. The discussion benefits from bringing implicit value-judgements more into the open.

---

[87] U.S. Preventive Services Task Force (2008); see also American Cancer Society Guidelines for the Early Detection of Cancer: http://www.cancer.org/Healthy/FindCancerEarly/CancerScreeningGuidelines/american-cancer-society-guidelines-for-the-early-detection-of-cancer?sitearea=PED

Chapter 6
Serving Society or Serving the Patient?

6.1 Summary of the Analysis so Far

The subject of the ethics of screening is vast and complex. It gives rise to several layers of issues, from the concrete level of actually conducting a screening programme, via complicated scientific and technical details connected to issues regarding the ethical assessment of risks and the goals of screening, to overarching policy issues regarding the basic criteria for when and which screening programmes should actually be conducted and what room for informed consent they should provide. Nevertheless, some general thematic threads can be discerned through the analyses undertaken in the preceding chapters.

First, many of the potential ethical conflicts or underlying value issues brought to the fore by screening programmes connect to the tension between applying a standard health care ethical or a public health ethical perspective on screening. In particular, prenatal screening programmes seem very difficult to justify unless their goals are formulated solely in terms of reducing the incidence of inborn disease in the population, while standard health care ethical requirements of respecting autonomy are discounted. A related example is the question of how to trade off severity and prevalence of the targeted disease when these factors pull in opposite directions. At the same time, the intricate problems surrounding the issue of what should be required by tests and follow-up procedures/treatments give rise to a need for value judgements that seem impossible to import from either of the perspectives in isolation, but rather create a need for mixing the two. It remains to be described, however, how such a mixing might be accomplished in light of the obvious tensions between these perspectives. Generally, it seems as though the intuitive plausibility of applying one or the other of the health care ethical or the public health ethical perspective varies, depending on what area of screening is discussed. For instance, in the prenatal area, an exclusive public health ethical perspective would seem to have implications that, for good historical reasons, give most people the proverbial creeps. In childhood and adult screening, or screening as a part of communicable disease management, in contrast, such a perspective may appear more acceptable to most people – especially if the situation is particularly serious, such as a full-blown pandemic or other sort of widespread ill-health that threatens to undermine basic societal infrastructure and institutions. At the same time, it may be assumed that many people would still be reluctant to abandon the health care ethical perspective entirely, providing as it does special safeguards for individuals against abuse by institutions and society. On this note, it may be observed that most childhood and adult screening programmes actually running seem to be examples of when the health care ethical and the public health ethical perspectives pull in the same direction. Besides being well motivated on the basis of their overall effects on population health, they achieve this effect mainly through providing participating individuals with obvious health benefits, e.g. hearing screening in early childhood. When such programmes become controversial it is mainly due to either doubts about their ability to deliver such benefits to an extent that justifies the costs of the programme, or about external factors (such as stigmatisation or discrimination) counteracting the programme's health promoting potential.[1] Both these sources of doubt are valid from both the health care ethical and the public health ethical perspective.

Second, all screening programmes give rise to a need for considering a much wider context of factors than that of medical facts about conditions, treatments, testing methods, the individual testing situation and the immediate practicalities of running a programme. Screening as an activity is deeply entangled in complicated social and societal webs of circumstances affecting the prospects of individual people as well as the overall outcomes of programmes in terms of population health. Several examples have been given throughout the analysis of how social and societal factors may undermine both patient autonomy and what would otherwise look as promising outcomes in terms of health. We have also presented a number of cases illustrating what was held out at the outset of this book as one of the chief reasons for studying the ethics of screening; the undeniable fact that there is a sort of business side to screening activities (also when they occur within the public sector) that threatens the credibility of reasons put forward by hospitals, centres or professional associations in favour of particular screening programmes. To spot and react to such threats, proposed

---

[1] A good example of a case where both these factors seem to be present in making the programme controversial is ATD screening (see Section 3.3.2).

6.1 Summary of the Analysis so Far

screening programmes need to be analysed not only on the assumption of being well-meant suggestions by people or groups having only the best interest of patients or the population in mind, but also as the outcome of organisational processes in which the very same people and groups have vested interests of other sorts. Moreover, decisions on whether or not to launch a programme, or on how it should be designed, can many times be more the outcome of societal processes outside of health care having nothing to do with balancing the pros and cons from an ethical and medical point of view. Obvious factors are the way in which a country has chosen to design its overarching health policies, or how the commercial sector is regulated, e.g., regarding the rights of employers. Less obvious, perhaps, but not less important, are the marketing strategies of the medical technological industry, for which screening programmes are, of course, virtual wet dreams. Unfortunately, this last factor connects in an undesirable way to the earlier mentioned factor of health professionals making decisions also on other grounds than the health and autonomy of their patients.

Third, many existing and suggested screening programmes seem to be based on a rather shaky ethical basis. This is partly a result of the factors already outlined, but also partly due to basic obscurities regarding what the goals of these programmes are supposed to be, in combination with both the latent conflicts between different goals and the different constraints implied by these goals (e.g., that of having an acceptable cost-benefit profile, or that of implementing an acceptable consideration of personal autonomy). In addition, both the confusion of different potentially incompatible goals and the adoption of different such goals in different areas of screening give rise to a basic question that have not been answered: On what basis should the ethical plausibility of the goals of screening programmes be assessed? Answering that question seems to us crucial, both for underpinning stable ethical conclusions regarding the more problematic screening areas (such as prenatal screening, mammography screening, or the new expanded neonatal programmes), and for coming to grips with the current situation of apparent capriciousness in the introduction and the design of new screening programmes.

Obviously, this book is not the place to settle this issue. We do, however, hold it out as an urgent area for further research in both health care ethics and public health ethics. Sections 6.3, 6.4, and 6.5 of this final chapter are devoted to sketching the rudiments of a possible framework for conducting such research. In Section 6.6 we will return to what bearing the result of the book has to the classic Wilson and Jungner criteria for screening. In Section 6.7 we will then end this book with some closing reflections. Before that, however, we will first revisit what was already at the outset held out as a major source of the ethical issues raised by screening.

6.2 The Public Health – Health Care Tension Area

The basic ethical tension between health care and public health has been increasingly commented on since about a decade.[2] The tensions can be located to particular areas of practice, as well as to basic assumptions and customary accepted lines of reasoning. A basic tension has been argued to be one of *perspective*: while medical and health care ethical ideas and arguments take their start in the meeting between a health care professional and a patient, public health ethics rather starts off from the interactions between institutions and populations.[3] Thus, from the very outset, the questions asked and the contexts assumed for the attempts to answer these questions are rather different. At the same time, it is not entirely clear exactly how the difference of perspective is to be characterised, since the relation between an individualistic and a population-based outlook can be construed in different ways that affect both the nature and the severity of the tension involved.[4]

Nevertheless, there are two areas where the difference of perspective can be assumed to stand out no matter how this difference is theoretically modelled. First, the view on what may count as a value or a good is restricted, within standard approaches to medical and health care ethics, to health effects on particular individuals. This idea is also applied in regulation, such as the Helsinki Declaration[5] and the CIOMS ethics code,[6] and echoed in the ethics codes of national medical associations and societies of health professionals in more or less all countries. Second, the room for trading off such individual interests against aggregated or "public" interests of the sort that is in focus in public health is highly restricted within standard models of medical and health care ethics. Third, standard medical and health care ethics does not provide much room for an idea of promoting patient autonomy that may justify that patient autonomy is not respected or, at least, that it is somewhat restricted. As we have seen, various screening programmes may appear troublesome in all these respects. At the same time, as was mentioned at the outset of this book, it seems hard to hold out a credible medical

---

[2] Among the earlier takes on the subject being Childress et al. (2002), while a recent major publication is Dawson (2011).

[3] Munthe (2008).

[4] For a recent discussion related to how different parts of public health differ from each other and interact with differing societal sectors and concerns, see Wilson (2009). For a broad outline of the different ways in which the population-focus of public health may be modelled, see Shickle et al. (2007).

[5] World Medical Association (1964–2008).

[6] CIOMS (2002).

or health care ethical idea that does not pay *some* attention to the importance of public health as a prerequisite for having a decent health care system in the first place.

The rather principled stances of standard medical and health care ethics are, at least at first glance, difficult to reconcile with the population-focused and more pragmatic viewpoint of public health. In the latter case, public and health related interests serve as guiding principles that allow for the interests of single individuals to be set aside in case of conflict. At the same time, such public interests are usually understood to include a certain amount of consideration for the conditions of individual people, as well as moderation in order not to undermine public trusts. This, for instance, seems to be the levers employed in Friedman Ross' case for accepting parental informed consent in neonatal screening.[7]

In spite of such openings for narrowing the gap between the health care and the public health ethical perspectives, the most common *organisational* approach to the many potential conflicts is to try keeping them apart. Sectorisation is the most common strategy. Vertically, the overarching public health considerations that need to be attended to for the health care system to remain operational are traditionally left to the political sector of (national) overarching policy making, while the actual operation and overseeing of this system is left to professionals, more or less autonomous public agencies and the legal system. Horizontally, day-to-day health care activities are normally occurring within institutions and settings kept separate from the operations of agencies and bodies responsible for implementing public health policies. At the same time, though, there is a significant deal of overlap between the sectors, not least horizontally.[8] Screening programmes, we hold, constitute one example of such horizontal overlaps, but so do, e.g., vaccination programmes, communicable disease management, health information campaigns, medically based nutritional interventions regarding, e.g. obesity and oral health, and so on. Not surprisingly, all of these examples also awake continuous controversy in society as well as professional and research-related debates.

---

[7] Friedman Ross (2011).

[8] The amount of vertical overlap depends to a great extent on the organisation in different countries of the policy making, the enforcement of regulation of various areas and the implementation of policy in other respects. For instance, while in some countries direct political intervention would be expected in case of some undesirable activities undertaken by an agency or institution (such as a hospital), in other national settings, courts and prosecutors would rather be seen as the principal responders.

In the case of screening, what has transpired is that this controversy has not produced a strengthened theoretical basis for the justification of screening, but rather sustained or even increased lack of clarity and systematisation and, as a practical result, substantial portions of arbitrariness. Possibly, this may be explained by the apparent fact that creating a clear and consistent ethical basis for screening would necessitate some rather uncomfortable decisions at the higher policy making levels of society – our discussions about neonatal screening, prenatal screening and mammography screening provide cases in point. Nevertheless, the arbitrariness provides room for screening programmes to be influenced by individual and institutional interests that should not have a bearing on health care public health activities, no matter what perspective on the ethics of screening that is taken. This observation takes us over to the next general point to consider.

6.3 The Relevance of a Social Science Perspective

Just as standard approaches to medical and health care ethics normally shun away from values attached to social considerations and entities, public health and traditional ethical reflection connected to that tends to portion off the concrete effects on single individuals of population based measures and policies. As always, there are exceptions to this general rule, such as the retrospective debates about the eugenic policies of liberal democracies in the twentieth century, discussion and research on the ethical implications of gene and reproductive technologies, and recent discussion on global health care and research ethics.[9] All of these exceptions exemplify a need to consider complex social and societal facts, mechanisms and processes that we have found also in the case of the ethics of screening.

All of these cases suggest that the importance of taking a social science perspective when discussing the ethics of activities occurring in areas of overlap between the health care and public health sectors can hardly be overestimated. It is this perspective that, in the case of screening, allows us to spot such things as the influence of the pragmatics of organisations and institutions, the various ways in which overarching social patterns (such as prejudice) or societal designs regarding, e.g., the relation between health care, business and law, affect the expected outcome of screening programmes, or how health economic assessments of such programmes are affected by taking requirements of autonomy, informed consent and

---

[9] See Buchanan et al. (2000); Glover (2006); and Kottow (2002), respectively, for good overviews over the three discussions.

6.3 The Relevance of a Social Science Perspective

counselling seriously. It is the social sciences that can educate and inform us about the workings and intricacies of these and other sort of mechanisms and processes of relevance. Without that information we will be much less prepared to spot how they interplay with ethics considerations, undercut the attainment of desirable results, or illegitimately exploit the lack of complete and coherent ethical foundations.

To be true, in mainstream medical and health care ethical research and debate, social and societal phenomena are indeed considered from time to time. However, with the exceptions mentioned earlier, they are then mainly used as a sort of backdrop for presenting a problem area, or considered as contributing to some phenomenon where the ethical foundation or verdict is not problematised. What we are suggesting here is that the social science perspective needs to be fed into the very discussion of *what makes for a good ethical foundation and justified ethical verdicts* regarding sectorially overlapping activities, such as screening, and this for two reasons. First, the assessment of the relative importance of the functions of different societal sectors needs to consider how these functions may interact with each other and other sectors in view of some overarching ideal of a good society.[10] Second, in order to appreciate not only how this interaction should be from an ideal perspective, but how it may be expected to be in the real world, we need to understand the actual workings of activities at this level of things. Such understandings must be incorporated into any suggestion regarding the ethical foundation of a particular sector, as well as for activities overlapping different sectors, and can be provided only by the social sciences. In order to substantiate this suggestion further, in the next and final section, we will sketch a framework or approach for accomplishing such an end and discuss what may be achieved through applying it in medical and health care ethics research.

The claim regarding the relevance of the social science perspective has, so far, only concerned areas of activity overlapping the health care and public health sectors. However, when taking on the spectacles of the social scientist and considering what this overlap consists in and emanates from, there appears to be good reasons for the claim that most (if not all) health care activities actually involve such and similar horizontal sectorial overlaps. We have already mentioned the obvious connection between the situation in a country in terms of population health and this same country's ability to provide cost-effective health care services. We may now add that this ability will be obviously dependent on what ethical aims are pursued and what

---

[10] An argument to this effect with particular regard to health care and public health can be found in Coggon (2010).

restrictions are applied. Something similar can be said regarding the connection between the health care and economic sectors – so famously depicted by Michael Walzer as residing on near to opposite ends of an imagined scale representing the varieties of values pursued in society.[11] Even a poor country with vast unattended health needs may be able to run a sustainable health care system if only the aims and ethical restrictions are (at least initially) designed to allow that – for instance, by allowing economic considerations to substantially influence health policy making, and by tolerating substantial individual adverse outcomes in terms of health as well as autonomy. Of course, such a system may be found ethically indefensible, but then the upshot has to be accepted that such a country will have to do without a health care system if it is to secure ethical defensibility. In more developed countries, we can see that allowances of this type are operating in a less draconian fashion, for instance, in the way priority settings, decisions on diagnoses and indications, and allocation of research funds are made on grounds incorporating considerations regarding, e.g., long-term social stability, the functionality of the work market, the needs of the pharmaceutical and medical technological industry, and so on.

In effect, the framework and approach presented in the next and final section has a potential for being relevant not only for the activity areas *saliently* overlapping the health care and public health sectors, like screening. There is, we have now suggested, a significant portion of sectorial overlap (and not only with the public health sector) going on throughout the entirety of health care and medicine. Therefore, health care and medical ethics need additional theoretical tools to systematically incorporate the sort of social science perspective we have argued is needed for fruitful ethics research applied to sectorially overlapping activities.

This concerns also public health, where the connections between minding the health of the population and, e.g., overarching economic or ideological objectives of society are well-known to connect and interact since before – quite apart from the obvious connection between the pursuit of population health and the operations of health care. Another obvious overlap with regard to public health is that with the criminal legal system, e.g., in the case of communicable disease management, the societal handling of mentally disordered offenders[12] or, as we have seen is a matter of debate, the possible denial of applying informed consent standards in some screening programmes.

---

[11] Walzer (1983).
[12] Nilsson et al. (2009).

6.4 An Institutional Approach to Health-Related Ethics: A Sketch

The framework we propose is built on the basic idea of looking at screening programmes not merely as a set of activities undertaken by a collection of health care and laboratory staff in an area of overlap between different societal sectors, but as actual *social institutions*. The logic behind this proposal runs roughly as follows: As a rule, screening programmes are quite autonomous organisational bodies within the health care and public health systems or particular health care institutions (such as hospitals), often regulated and governed in specially designed ways. As such bodies, they are in various ways related to other institutions of society; research institutions providing relevant knowledge and technical skill, health institutions providing treatments and follow-up procedures, political institutions setting health policy agendas as well as relevant regulations, public health agencies overseeing and directing the implementation of such policies and informing the public about this, private enterprises selling tests, the legal system handling misconduct of any of the named parties or individual citizens and media absorbing, interpreting, and reporting on public and other reactions towards screening programmes. Thus, by framing the outlook on screening programmes in terms of an institutional approach, ethics immediately gains access to an analytical scheme both making visible the sort of sectorial overlaps highlighted earlier, and being easy to relate to social science information on how politics, organisations and institutions function and tend to interact.

To apply this sort of approach means that the ethical assessment of, e.g., screening programmes must concentrate less on the actions of individual health care professionals, since these are all figurative pawns in the larger games played by institutions of this kind. Earlier in this chapter, we have suggested that this holds also for several other (and perhaps all) health care and public health activities. This is not to say, of course, that ethical assessment of the actions of individual professionals is to be taken out of health care or public health ethics – such an idea would border on the absurd. However, we do suggest, that such investigations need as a rule to be anchored in an ethical analysis of the institution in which such professionals are operating. Thus, when considering from an ethical point of view the actions of a doctor or a nurse interacting with patients in a screening context, it may be that, due to institutional considerations, things that would in other settings be held out as either problematic or innocent, become permissible or rather problematic when occurring within a screening institution. For example, in this book, we have pressed the point that, due to the significant power imbalance between a screening programme and a single person, attempts to even mildly persuade patients to enter a screening programme is extra problematic in terms of autonomy compared to other situations where a doctor and a patient discuss the pros and cons of a taking a certain test. This argument can now be further backed up by pointing out how the doctor in the one case is acting primarily as a representative of a societal institution approaching a single individual with proposals from society about consuming certain health services and, in the other, more as an individual professional provider of services responding to a request from a patient that turns to society for assistance.

Approaching screening programmes as institutions also provide conceptual tools for working on those ethical issues earlier held out as most problematic. First, if each screening programme is seen as an institution, the phenomenon of different programmes pursuing different goals need not come out as a devastating problem. The differences between programmes may be justified by analyses demonstrating how different programmes play differing institutional roles (that is, filling different desirable functions) in a good society. However, such analyses may also reveal that certain (existing or proposed) programmes do in fact not play any such role – at least not one that can be substantiated as desirable.

In a similar vein, it may be possible to find some sense in the apparently arbitrary variation as to how different ethical perspectives and values are mixed in different programmes when these are understood as institutions, each of which is constructed to perform a particular function in society. Just as before, it is an open question to what extent the mixes found in different programmes can in fact be justified. To investigate that, a basic analysis of the institutional role of the particular programme is a necessary starting point.

Another important upshot of the institutional approach is that when screening programmes are perceived as particular institutions guided by specific goals, embodying special mixes of ethical perspectives and values, and performing specialised societal functions, the notion of *institutional integrity* becomes applicable. Already proposed as a key concept for exploring the limits of legitimate health care activity, as well as for applying organisation ethical theories and solutions to the medical and health care setting,[13] this notion can help us understand what is bad about several phenomena encountered throughout the analysis in the foregoing chapters. This regards both the "leakage" of values that seems to be going on between some screening programmes and closely related societal sectors, and the concrete influence of the occurrence, design and development of screening programmes from

---

[13] See, e.g., Smith Iltis (2001).

outside parties or forces that have been held out as problematic. This regards, for instance, the way in which the medical technological industry often seems to be allowed to set the agenda for prenatal screening programmes, or the way in which neonatal screening programmes are redesigned for the sole purpose of avoiding lawsuits, while health care or public health ethical reasons rather hold out the changes as problematic.

To round up this brief sketch, it should be emphasised that the institutional approach is not suggested as a *replacement* of standard approaches found in either health care ethics or public health ethics. Rather, what we suggest is that this kind of approach can be useful as a complementary perspective within both of these areas, in particular with regard to activities occurring in areas of overlap between them. We are also proposing that this approach would be valuable for ethical analysis of other areas dealing with the health of individuals or the population, where there is close interaction with other societal sectors (which we have suggested to be the case with not only screening programmes, but most health care activities). Of course, this usefulness of the institutional approach can only be thoroughly demonstrated through actual applications with successful outcomes. The analysis of the ethics of screening undertaken so far seems to provide some prima facie reasons in favour of giving our suggestion a try. We will now try to provide some further evidence by way of three application cases that we hope may further inspire researchers in applied ethics and health research to attempt more in-depth investigations of the kind we are suggesting in specific areas.

6.5 Applying the Institutional Approach: Three Cases

In order to have some more concrete illustrations of how we envisage the institutional approach to work – and also illustrate some inherent problems that need to be tackled whenever applying this approach – we will briefly consider three screening-related cases. One of these has already been highlighted repeatedly in this book: the striking difference between our apparent moral reactions to the issue of how the health care ethical and the public health ethical perspectives should be mixed in the ethical analysis of screening in the communicable disease and the prenatal areas, respectively. The second case is brought in to provide an illustration of what was said in the first chapter about how the ethical issues and problems of screening may very well reappear in health related activities that are not *bona fide* screening programmes. We will, therefore, briefly consider the currently debated case of direct to consumer (DTC) genetic testing. As a third and closing case, we return to the issue highlighted in Chapter 2 as the main one where considerations of justice are central to screening: the question of how screening programmes should be assessed from the point of view of rationing scarce health care resources.

6.5.1 Institutions, Functions and Ethics: Prenatal Care vs. Communicable Disease

The difference between our spontaneous moral reactions to what goals and ethical restrictions are appropriate in the respective fields of prenatal screening and screening in the context of communicable disease management have been highlighted in several places in earlier chapters. We characterised the difference partly in terms of what mix of considerations of individual and population health should be seen as part of the goals of screening in these two areas, partly in terms of what restrictions in terms of autonomy that may seem acceptable.

In short, in the communicable disease context, the idea of having as the main goal of a screening programme to promote the overarching health of the population would seem to be acceptable. Thus, in certain imaginary cases, screening may be well motivated mainly due to the expected prevalence of a disease and, in pressing such situations, many of us would accept the use of coercive means to make the screening contribute effectively to, e.g., the containment of a threatening major pandemic. Although we, of course, would prefer if such conditions did not arise in the first place, if they still did, we recognise the possibility that extraordinary measures may be motivated.

Applying similar reasoning to prenatal screening, however, seems to provoke quite a bit of moral unease. This regards, in the first instance, the idea of making screening obligatory or applying less direct coercive means to have pregnant women join prenatal screening programmes. A focus on population health, rather than individual health, in the design of prenatal screening is the most straightforward rationale for the application of less than full respect for reproductive autonomy. As we have seen, the moral unease with such a focus creates a basis for a lot of the criticism that has been wielded against prenatal screening – e.g., with regard to its structurally conveyed devaluating messages about disabled people, the ongoing shift from selecting target conditions mainly on grounds of severity to letting prevalence play a larger role, and the very idea of organising prenatal testing in the form of screening programmes in the first place.

Our reason for highlighting this difference as to what basic values and moral standards we are prone to apply to screening has been to illustrate

6.5 Applying the Institutional Approach: Three Cases

the extent to which the actual practice of screening as a whole seems to lack a coherent ethical foundation. However, in this chapter, we have also held out the institutional approach as a way of deepening our understanding in a way that may reduce such incoherence – or, at least, point to ways of reducing it. The basics of doing that lies in the task of looking at screening programmes in the contexts of communicable disease management and prenatal (and reproductive) medicine as separate institutions, the point of which are to fulfil different societal functions. So this is what we will do rather briefly, only in order to illustrate how a pattern of reasoning may proceed from that basis.

Communicable disease management is one of the classic core areas of basic public health work. It has a long history in human societies and, looking over this background,[14] it appears to have been instigated for fairly consistent reasons throughout the centuries, albeit the means at its disposal and its actual rate of success has changed over time, not least due to the breakthrough of biomedical science in the late nineteenth century. In most countries, there are special regulations (or special provisions in general health-related laws) governing the actions of agencies and professionals involved in communicable disease management. The function of this societal institution is fairly clear and straightforward: to prevent, contain and reduce the overall societal impact of epidemic or pandemic outbreaks of infectious disease. The goals pursued are thus firmly rooted in traditional public health ideology, where population health is in focus with a keen eye at maintaining basic social stability (involving public order, economy and basic infrastructure).

Accordingly, a screening programme in this context has as its desired function to contribute to such goals. This implies, for instance, that screening for diseases with low potential prevalence is difficult to motivate. Likewise, screening for a very contagious disease cannot be motivated merely by pointing to a possibility of curing a fraction of those tested positively. Screening within the context of communicable disease management thus seems to need to meet three conditions: One, the disease constitutes a sufficiently severe (potential) threat to population health and/or social stability. Two, the disease can be reliably detected by a screening. Three, there is a feasible follow-up procedure that, if applied to positively tested people, would effectively prevent or sufficiently reduce the threat to population health and/or social stability. As mentioned, implementing such a screening programme may involve coercive measures to ensure a sufficiently good uptake, as well as when applying the follow-up procedures. As long as this sort of possibility

---

[14] Rosen (1993).

is not abused or overused, it seems to be well in line with the function that society as well as most individuals expect communicable disease management to perform. From a normative standpoint, it also looks as a perfectly justifiable function to have in a society – more or less regardless of what basic ethical views you may happen to entertain.

If we now shift our gaze towards prenatal and reproductive care, we may notice certain familial relations to the communicable disease field as regards the historical background. Historically, prenatal and reproductive care has been an important part of public health, connected to the longstanding interest of nations and societies in the composition, size and functionality of the population.[15] However, as modernisation has proceeded, wealth has been accumulated and the health of the population has increased, more and more power has been transferred to individuals regarding reproductive matters.

Although there are some notable ideological overlaps,[16] there is a tremendous difference between the values driving the eugenic policies of almost the entire developed world during the first two thirds of the twentieth century and those advanced as central in genetic counselling connected to prenatal diagnosis since the mid 1970s. Even more salient is the general loosening of society's grip on the reproductive decisions of individual people occurring during roughly the same period. Today, while reproductive care is certainly seen as performing an important task from a public health perspective, it is at least just as much seen as an institution whose function is to serve the interests and values of individual people. In effect, while most people would probably acknowledge the importance of having society score well on the classic public health variables of infant morbidity/mortality and the morbidity/mortality of pregnant women and new mothers (both highly sensitive to variations as regards the quality of reproductive care), they would not agree to have that objective motivate much of coercion or manipulation of pregnant women or new mothers.

What is accepted, though, is some coercion in the name of the best interest of the child. Although the issue has sometimes been debated,[17] in practice, this room for restricting the autonomy of pregnant women has, however, as a general rule been limited to cases after or during birth. Except for the case discussed in Chapter 3 about parental consent in neonatal screening, such provisions have little to do with public health considerations, however.

---

[15] Indeed, one of the classical findings of infectious disease and how to prevent it to spread (Semmelweis' famous testing of his bacterial infection hypothesis regarding puerperal fever) comes from the field of reproductive care (Hempel, 1967).

[16] Munthe (1996).

[17] Coutts (1990).

Another general trend is that the times when public health considerations were allowed to exert a substantial influence on reproductive care (e.g. in the form of programmes of compulsory sterilisation) are nowadays looked back upon as a barbarism very far from what we expect reproductive care to deliver. In Chapter 3, we considered whether, in spite of this, there may be situations where developing countries allow such thinking as a transitionary phase on their way to a better public health situation. Besides eugenics, this might include as examples the "pro-natalist" policies of Romania in the 1970s and 1980s, where fertile women where forced to undergo monthly gynaecological examinations (which, of course, is a case of screening) in order to make sure that they were not robbing society of any future citizens by having an abortion.[18] We concluded, however, with great scepticism to such prospects and, indeed, the eventual collapse of the Romanian attempt may be seen as providing a grim case in point.

Applied to prenatal screening, we can now see why it makes sense to set quite different goals and ethical restrictions for this practice as compared to screening for communicable disease. Prenatal screening is an institution that is in place to do a quite different (albeit in some minor respects related) job for society than screening in the service of communicable disease management. Possibly, this reflects an overall change in people's views on what a decent or good society amounts to. Hence the differences with regard to what is seen as the desirable function of reproductive care today compared to 70 or 100 years back. This, in turn, means that an analysis made by applying an institutional approach seems to underwrite the criticism that we have wielded against the way in which actual prenatal screening is run. The function of the institution of reproductive care would be served better by abandoning the screening approach, while still making prenatal diagnosis available after proper counselling at the initiative of interested women and couples.

6.5.2 Direct to Consumer Genetic Testing: The Limits of Context Relativity

The last decade, so-called direct to consumer (or DTC) genetic testing has been a growing business coming out of the success of genetic research and technology development. DTC genetic testing is simply the commercial sale of genetic testing services without necessarily involving medical specialists

---

[18] Kligman (1998).

or genetic counsellors in the process. Technological development has led to the situation where scenarios sketched and problematised by ethicists 15–30 years back are now very much a reality. The first examples (from the 1990s) were provided by commercial IVF clinics promising healthy babies of the right gender by including preimplantation genetic diagnosis as a part of the routine Now, we have companies providing small testing kits to pharmacies or other retailers to sell over the counter or directly to consumers over the internet. An intermediate version is the offering from commercial genetic labs of testing for a large number of genetic variations known or claimed to be of importance from a health perspective, which has been going on since the early 21st century. And the technological development promises more in this vein, as has recently been highlighted in the medical research literature[19]: more genetic variants and more claims as to their relevance for the health minding person.

The reason why we have chosen to highlight this particular area is that it is a good example of a practice of using, offering and organising medical testing in a way that is, strictly speaking, *not* screening, but that still shares enough of similarities with screening for many of the arguments and points made earlier to be applicable. DTC genetic testing is furthermore interesting since it is an area pursued from a sectorial angle (business) where the goals and values pursued are very far from those driving typical screening programmes. This presents a challenge to the institutional approach. As was seen in the preceding subsection, what is the (proper) function of an institution or societal sector may change with the context. Does this mean that the institutional approach is doomed to fall into the abyss of mindless – as well as useless – relativism? We hope to demonstrate by considering the example of DTC genetic testing that it does not. While context sensitivity of an ethical analysis will always mean that ethical conclusions regarding different phenomena will be somewhat relative to circumstances and context, the institutional approach provides limits for how deep and wide such relativity can become.

DTC genetic testing is an example of medical tests being offered at the initiative of someone else than the patient to great masses of people, where the latter are not (necessarily) united by a previously recognised risk or worry for which the test is relevant. Moreover, the function of DTC genetic testing is to be a filter for subsequent decisions on various follow-up measures (to be decided on by those who purchase a test). All of these features are essential constituents of the basic concept of screening applied in this

---

[19] Evans et al. (2010); and Annes et al. (2010).

6.5 Applying the Institutional Approach: Three Cases 143

book and explained in Chapter 1. The crucial difference between DTC genetic testing and the sort of screening programmes that have been in focus earlier in this book is the identity of the party taking the original initiative. Rather than society or the institution of health care, in the case of DTC genetic testing, the initiative is coming from the commercial sector and is, clearly, based mainly on business considerations. Moreover, there is no specific predefined population to which the offer of testing is directed (unlike, e.g. all newborns in PKU-screening). Rather, anyone that might develop an interest in purchasing the test as a result of the marketing efforts of the company offering it is part of the population targeted by the initiative taken by this company.

One could, of course, let this fact be the start of a semantic discussion as to whether or not this difference means that DTC genetic testing is not screening. That approach, however, would lead nowhere, or so we argue. For even if we assume DTC genetic testing not to be screening just because it is coming from a pure business perspective, it is still true that most of the ethically relevant features and considerations taken up throughout this book are applicable. The recognition of this has recently provoked intense debate in the US[20] and made the Food and Drug Administration halt the planned introduction of genetic testing kits to be sold over the counter in the shops of a large pharmacy firm pending assessment and authorisation.[21] Similarly, in the UK, the Nuffield Council of Bioethics has, at the time when we are writing this, just released a report noting that all the problems of genetic screening (as well as regular testing) seem to reappear in the DTC genetic testing area.[22]

The line taken by the FDA in the just mentioned case, as well as the Nuffield Council, rests on the idea of the service offered by the companies in question to fall under regulation or ethical principles aimed at guaranteeing a certain standard of quality and safety of medical products. However, this does not – in the case of FDA – necessarily imply that an ethics of screening should be used as a yardstick before deciding on a possible authorisation of the DTC-product, even though the product has similarities to screening proper. Similarly, while the point of the Nuffield Council that the mentioned ethical issues are applicable may be noted, the conclusion that the particular practice of DTC genetic testing therefore needs to conform to certain standards may be resisted. So, simply put, the question is if defensible DTC

---

[20] Stein (2010a).
[21] Stein (2010b).
[22] The Nuffield Council on Bioethics (2010), chapter 9.

genetic testing activities *should* conform to standards coming out of the considerations made in connection to screening (in turn implying that these standards should play a part in the assessments made by agencies such as the FDA). What would the institutional approach have to say about this?

Answering this question quickly leads to a critical issue in interpreting what an institutional approach to health-related ethics amounts to. For, apparently, an argument along the following lines would seem perfectly possible to launch: DTC genetic tests are products sold for business reasons by commercial companies and purchased at a monetary price by customers apparently judging the value of the product to be worth the price. In this description we could exchange the phrase "DTC genetic tests" for, e.g. "loafs of bread", "pants", "bicycles", "houses", "movie tickets" or any other type of consumer goods. To make the analogy closer, we may compare the product of a DTC genetic test with other available products the value of which comes from providing the consumer with certain information; newspapers, wire services, credit rating, value assessment of equity, and so on. Thus, DTC genetic testing is just another product bought and sold in the societal sector of business, at the initiative of a commercial company. Now, the primary goal of commercial companies is simply to make money and profit, and the only constraint here comes from what may undermine the making of profit. Similarly, it is a rather common idea that the sector of business taken as a whole has the function of doing exactly that: accumulate profit and wealth. Besides Michael Walzer,[23] mentioned earlier, we can also note the classic statement of Milton Friedman that "[t]here is one and only one social responsibility of business – to use it resources and engage in activities designed to increase its profits so long as it stays within the rules of the game, which is to say, engages in open and free competition without deception or fraud",[24] aimed at the notion of having commercial companies within a competitive market system assuming any more demanding responsibilities than making money for their owners. The upshot of such an analysis would then seem to imply that the many features and aspects of the ethics of screening taken up earlier, as a matter of fact, have no necessary bearing on the case of DTC genetic testing. As long as the sort of problems and downsides we have been taken up as sources of ethical problems in screening tend to negatively affect the profit margins of the companies selling DTC genetic tests, they provide no reason for these companies to moderate their activities or for society do to that for them.

---

[23] Walzer (1983).
[24] Friedman (1970).

6.5 Applying the Institutional Approach: Three Cases

Such an analysis, however, is incomplete in several respects. First, mismanaging DTC genetic testing from the point of view of a health care or public health ethics perspective may not threaten the profit margins of individual companies, while still undermining the growth of the entire business area of DTC genetic testing. Individual companies can be started, dismantled and renamed swiftly enough for the investors to rescue their business, should the activities of one company stain its credibility. However, widely occurring mismanagement may still create public distrust in the whole area, spilling over also to serious providers of DTC genetic testing – and even genetic medicine in general, including neonatal screening, for instance. Second, a crucial condition of Friedman's statement is that of the activities of DTC genetic testing providers taking place "without deception or fraud". However, the following two facts comprise the main reasons for why DTC genetic testing can be such good business: (a) it provides (very often complex and ambiguous) genetic information without the involvement (or sometimes less involvement) of intermediary medical or counselling expertise (especially regarding pre-testing counselling), (b) DTC genetic testing companies employ the tools of marketing to attract customers. Due to these facts, DTC genetic testing is vulnerable to serious criticism also from the point of view of Friedman. Let us elaborate.

Given the severe problems mentioned earlier in this book of making people aware of the many aspects of genetic information of relevance to the quality and plans of their lives, the dodging of the steps of medical consultation and proper genetic counselling means that it is very likely that many customers would purchase tests without having been made sufficiently aware of the possible consequences for themselves, their family and their offspring. If we add to this the way in which the tools of marketing are in fact used in this area, the label "deception or fraud" comes very close to what is actually taking place. For instance, in the DTC genetic testing business connected to commercial IVF, being met by seductive images of adorable babies and happy and materially successful families is legio, as is a host of other emotionally loaded signs of social success, and statements hinting that genetic testing is a given step in the process of forming a family, that it is needed to build a healthy family, and so on.[25]

DTC presymptomatic genetic testing provided by commercial laboratories cannot capitalise as strongly on the reproductive theme, but press hard

---

[25] The curious reader may inspect these and further ingredients in the marketing of commercial PGD via the information and links provided at the *Infertility Resources Website*. It may be added that the centres and clinics linked to there are far from the worst examples of aggressive and manipulative marketing in this field.

on the notion of people being given a chance to take control of their own health (thus implying those not doing that to be less caring or responsible regarding their health). In addition, it is customary to even the balance further by rather blatantly misrepresenting what reasons there are for taking the tests and what may be shown by these tests. One regularly used trick is to describe only the (most severe) symptoms of the diseases and say very little about the nature of the genetic link or the probabilities involved (or bury this information several clicks away on the webpage). Another trick is to focus on *relative* risk figures without making clear how these differ from absolute figures and even using a language and visual illustrations conveying the strong impression that the relative figures are in fact about the absolute risk of contracting a disease given the presence of some genetic mutation.[26]

The argument made so far only concerns the reasons from within the sector of business itself to actually apply many of the ethical requirements highlighted in earlier chapters to DTC genetic testing. We may press this point somewhat further: if a company mismanages its provision of DTC genetic testing in the just outlined ways and therefore acquires a stained reputation, the rational thing for the owners may be to have this particular company closed down and to start again with a fresh market-brand. But then we are no longer talking about an interests flowing out of an institution. Not business as a whole, for sure, but not even the individual company; we are talking about the vested interests of investors, who in the end boil down to individual people making money. At that point, we hold, the borderline sketched by Walzer and Friedman between the sector of business and the concerns of other societal sectors breaks down; what we have in the end is a case of one person making money out of the manipulation and deception of another person, possibly causing a number of negative effects for the latter.

This, by itself, is a reason for society to step in simply in virtue of its general duty to protect its members against the abuse of others. However, even if we refrain from pursuing the analysis that far, we may note that both Friedman's and Walzer's ideas about business as a sector with making money as its only function is perfectly compatible with the notion of a society that balances the good of this function against that of the functions of other sectors. That is, the institutional approach allows for the assessment of the overall importance of different sectors and the ways in which they perform their functions. A simple example is the way in which all nations constrain the sector of business in relation to possible ventures that would be

---

[26] See, for instance, the *DnaDirect* web portal, and the blatant example of the marketing of genetic testing for melanoma at the *Myriad Genetics* website.

a threat to the national security, social stability or – as a matter of fact – the health of the population and individuals. While the accumulation of money (or wealth) is, of course, important for any society, its importance from an institutional point of view may apparently be affected by how it affects other important societal sectors and institutions.

So, in the end, what ethical demands are to be applied to DTC genetic testing from an institutional point of view boils down to the issue of how a decent society would balance, on the one hand, the importance of the contribution to the national economy made possible by applying relaxed demands and, on the other, the importance of protecting people from the various downsides of screening. This, we hold, is a rather easy issue, no matter what more particular idea one may entertain about the features of a decent society. There is no reason not to apply exactly the same ethical requirements to DTC genetic testing as to genetic screening programmes run by hospitals and other public health institutions. If DTC genetic testing providers can meet these requirements and still make a profit, fine. If not, we may conclude that the very idea of a sustainable business of DTC genetic testing was a fantasy to begin with.

6.5.3 Screening and Justice: When to Spend Health Care Resources on Screening

In Chapter 2, we considered the idea of justice as a central goal of screening, concluding that such a notion is secondary to goals that are about whether or not that which will be provided by a screening programme is in fact any good. We also suggested that while the design of a screening programme raises many cases of having to set priorities, these can be dealt with without having to resort to any particular theory of justice. However, as pointed out in that context, there indeed seems to be one issue with regard to screening that is predominantly about justice. This is the issue of how much of the resources available for health care and/or public health that should be allocated to screening programmes, assuming that these programmes are ethically defensible in other respects. This topic is huge in itself, since it leads right into a general discussion about the just rationing of health care resources,[27] as well as overarching questions about how a society needs to be arranged in order to be just.[28] To no surprise, then, we will refrain from the

---

[27] See, e.g., Daniels (2007); and Persad et al. (2009).
[28] See, e.g., Rawls (1972); Barry (1991); and Lamont and Favor (2007).

attempt of ending this book by writing another one on these issues. Instead, we will limit ourselves to two more humble tasks: First, we will sketch very briefly how the institutional approach relates itself to this issue. Second, we will set out more systematically what findings made earlier in this book are relevant for the allocation issue and what this implies for decisions on this issue.

The institutional approach means that we look at screening programmes as institutions performing sectorised societal functions. One point of this outlook, highlighted earlier, is that the institutional approach thereby allows for some variation as to what is to be seen as the goal of different screening programmes. Another point, partly described in the preceding subsection, is that the functions and/or goals of different screening programmes can be comparatively assessed from the point of view of society as a whole. This is no different than when society assesses the relative importance of other societal sectors, such as business and finance, social security or law enforcement. This means that, as a matter of fact, the institutional approach is quite fitting for applying to the issue of just health-resource allocation to screening programmes. Again, assuming that a programme is not burdened by other ethical problems, from society's point of view, this issue will be about:

(a) how well a programme perform its function (i.e. attains its goal),
(b) the societal importance of this function,
(c) how much resources are required for the programme to perform its function, and
(d) how the programme scores comparatively relative to other health- or health care related activities that the resources may be spent on.

It is step (d) that requires the sort of general principles of justice that we will not venture to explore further. Such principles will make clear how the facts and relationships given by (a)–(c) determine the score mentioned in (d).

With regard to (a)–(c), however, the institutional approach has somewhat more to say. This connects partly to the social science perspective held out earlier as one of the strongholds of the institutional approach. One factor of importance made salient by this perspective is the various structural forces and mechanisms active in relation to (proposed) screening programmes creating a strong pull for introducing new such programmes and emphasising apparent reasons in favour of such introductions. Another factor that is perhaps less connected to the social science perspective, but nevertheless important, is the general outcome of the ethical discussion of earlier chapters in relation to the expected costs of screening. Simply put, if a screening programme is to have any chance of securing ethical defensibility in terms of health and autonomy, it can be expected to be quite expensive.

Taken together, these two factors (the pull towards screening and the costs of ethically defensible screening) makes it plausible for society to apply a reversed burden of proof with regard to the issue of just resource allocation. The general principle should be that screening programmes are not to be allowed resources, unless it has been clearly demonstrated not only that they are ethically decent in their own right, but also that the necessary resources would not be better spent elsewhere in the health and health care sector.

The analysis of earlier chapters have revealed many existing or seriously contemplated screening programmes to suffer from rather striking ethical difficulties that would be quite costly to deal with – if, indeed, they can be dealt with in any other way than shutting the programme down. This situation, we suggest, is at least partly a result of society being much too allowing in its policies relating to the introduction of screening programmes.[29] The issue of justice would seem to add considerably to the importance of this finding, since it highlights the possibility of other important health or health care areas to be sacrificed as a result. The combination of a strict assessment standard and a reversed burden of proof is a first necessary step towards having society take control. After that, debates on what allocation would indeed be the just one may proceed, only now with a clearer view as to what one should be aiming for when making such decisions.

6.6 Revisiting the Wilson and Jungner Criteria for Screening

We have now said a great deal about the ethics of screening; its components, content, and practical implications. However, some would perhaps like even more concrete guidelines. Moreover, some would perhaps like to know how what we have been saying so far relates to the classic Wilson and Jungner criteria presented at the outset of this book. Addressing these concerns, we will argue that, in the light of our findings up to this point, the Wilson and Jungner criteria for defensible screening hold surprisingly well – in spite of them being more than 40 years old. The different points of a general and more specific nature made throughout this book do not undermine or invalidate these criteria, but rather add nuance and complication to both how they are to be understood and applied and to what is at stake in failing to apply

---

[29] At least partly, this overly allowing stance may be due to the way that screening programmes are often set up already at the stage of medical research, combined with the fact that most societies have a rather fumbling grip on the step from research to routine health care operations (partly because the last stages of medical research as a rule occur within the structure of clinical health care settings).

them. Properly interpreted, the Wilson and Jungner guidelines still function as a general frame for visualising what areas that need to be addressed in order to demonstrate a case of defensible screening. We will present the criteria and comment on them, suggesting in what way they are still plausible and how they should be interpreted more specifically.[30]

1. *The condition sought should be an important health problem.*

When Wilson and Jungner wrote this, they had a rather narrow conception of health in mind, almost equivalent to the notion of absence of disease, as understood from a strictly biomedical point of view.[31] Although this view of health is considered to be to too narrow today *in general*, and plausibly so,[32] it still seems appropriate *from the point of view of screening*. As we have seen throughout the book, various ways of broadening the application of screening have been attempted, e.g. by widening the scope of the concept of treatment (see Chapters 2 and 3) or benefit (see Section 5.2). One important upshot of these conceptual strategies has been to make the screening tool available for other purposes than addressing health problems narrowly conceived, e.g., for concerns of a psychosocial or autonomy-related nature. However, as we have repeatedly argued, although these purposes may be a very good argument for health care to offer medical investigations, tests or other services, they make a notably weak case for organising these offers in the form of screening programmes. On the contrary, reasons connected to considerations of autonomy and psychosocial well-being mainly seem to tell against such a way of organising the offering of tests in health care. Thus, Wilson's and Jungner's first criterion may still be understood as they themselves did at the time – the condition that a screening programme addresses should be a physiologically serious disease.

Of course, there is a remaining problem about determining exactly how serious the disease has to be, and this problem will need to be handled in connection to every single case. Moreover, seriousness can be understood in terms of prevalence or in terms of severity of suffering, disability and life-expectancy. As we have seen, different screening programmes understand seriousness in different ways. This, however, is not a problem in itself: to emphasise different goals within different programmes is well in line with ethical analyses on the basis of the institutional approach. What may become

---

[30] The criteria are quoted from Wilson and Jungner (1968), pp. 26–27.
[31] See Boorse (1997), or Wakefield (1992) for explications of biomedical concepts of disease.
[32] Nordenfelt (2003).

a problem is if the assessment of whether or not a (suggested) programme meets the criterion of addressing "an important health problem" with this in mind misses out on analysing adequately what values are central to pursue in the particular context of sectorial overlap actualised by this programme. We have argued prenatal screening to be a case in point.

2. *There should be an accepted treatment for patients with recognised disease.*

For Wilson and Jungner, this was the most important criterion. They understood it to mean that defensible screening programmes need to be able to offer a treatment that prevents, cures, or, at least, significantly ameliorates the disease targeted by a screening programme. That is, their view on treatment encapsulates their narrow view on what is to count as an important health problem. We have argued that, regarding screening, such a narrow view is indeed quite plausible: there should be an effective medical treatment in order for screening to be warranted.[33] To this we have added the idea of health promoting counselling as a treatment for those who test negatively as a factor that may add to the value of a screening programme, albeit not capable of justifying screening by itself. As noted, there are controversies about exactly *how* efficient a treatment has to be, what degree of evidence for its efficiency that needs to be present, and what amount and type of side-effects that are tolerable. Moreover, these factors have to be balanced against each other. We have already addressed these issues and will not repeat ourselves (see e.g. Sections 3.1.2 and 3.3). So, although the scope of the principle has to be made more precise (and we have indicated how), it is basically still a sound one.

3. *Facilities for diagnosis and treatment should be available.*

This criterion should be uncontroversial; it is hardly any point in implementing screening programmes if people are practically unable to access the benefits it may provide in theory. Moreover, facilities which are equally

---

[33] The hypothetical example of screening in the case of a serious pandemic where access to vaccines or cures are limited (see Section 2.2.2) does not depart from this example, although, in the worst case, it may be that the only measures available are those that aim at containing the infection at a population level by impeding people from moving about as they wish. In this case, what makes the health problem important is basically its potentially high prevalence and, as much as one would wish for sharper medical tools, temporarily restricting the freedom of movement of people is a biomedically certified way of managing this prevalence.

accessible for the *whole* population that would be benefited by screening are supported by considerations of justice (see Section 2.3 and Chapter 4). However, this criterion is, in itself, weak, since it says nothing about what should be required by the organisation of the facilities, e.g. in terms of adopting procedures of counselling and the collection of informed consent. Also this criterion, thus, needs to be made more precise, a task that we have embarked on (see Chapter 4). Although weak, the criterion is nevertheless useful to question some suggestions of screening programmes, suggestions that have a difficult time to explain how they should implement and organise screening in practice. Two obvious examples are neonatal or childhood screening for ATD (see Section 3.3.2) and neonatal screening for fragile X (see Section 5.2), but the problems regarding counselling seem to put into question also the more drastic expansions of general neonatal screening programmes having taken place in recent years.

4. *There should be a recognizable latent or early symptomatic stage.*

Historically, this criterion relates to the notion of screening as a *preventive* venture for the sake of *public health*. However, as we have seen through many examples, the advantage of being able to foresee and subsequently prevent a public health problem is not very different as regards ethical importance of doing the same in the case of a single individual. Moreover, the advantage of strict prevention is not *much* different from the advantage of being able to detect a disease in its very early stages and, therefore, being able to cure or ameliorate it more effectively. Indeed, also those reasons in terms of autonomy that may seem to tell in favour of screening point in the same direction. However, it still holds that if a disease cannot be detected before it is notably symptomatic, there is no advantage with organising a screening rather than offering regular health care examinations at the initiative of individuals, since when symptoms appear people generally seek health care anyway. This is the explanation why this criterion remains a reasonable one for *screening*.

Thus, this criterion (among others) casts serious doubts, for instance, over neonatal screening for fragile X, since it is unlikely to be a great difference in time between correct diagnosis with screening compared to correct diagnosis without screening. Moreover, and more importantly, difference in time will likely result in little difference in advantages (see Section 5.2), although there may be exceptions.

At the same time, we have also seen that the detection of a disease or a disease risk before symptoms appear bring a number of possible downsides. These have to do with what may be shown by the applied test and the subsequent effects on people in terms of both autonomy, well-being, and physiological health (e.g., due to overtreatment). In other words, this criterion must not be understood as a *carte blanche* for screening (given that other criteria are met). To make the criterion more justificatory forceful, it would have to be amended by the condition that the various downsides that may follow a presymptomatic test result is appropriately balanced by the benefits.

5. *There should be a suitable test or examination.*

Formulated in such general terms, this criterion cannot be questioned. However, there is room for substantial disagreement over what makes a test "suitable". We have discussed this at length (see Section 3.2). At this point, we can note that not only safety, validity, and predictive power are important factors in determining the suitability of a test. For example, identifying *true* positives can be problematic when it means overdiagnosis and overtreatment (see Sections 5.3 and 5.4).

6. *The test should be acceptable to the population.*

Above, the point was made that none of the Wilson and Jungner criteria can be seen as sufficient for warranting screening programmes. At best, each criterion describes a necessary condition. It is especially important to emphasise this elementary point regarding this criterion, since it has become all too common to defend screening programmes with reference to the alleged presence of a popular demand.[34] In reality this means, at best, "what the majority wants" (which, of course, is not the same thing as "what each individual want"), but more often "what some influential patient groups advocacies want". If one really cares about what people *want*, there is no substitute for high standards of informed consent and counselling, standards that screening programmes have a hard time to fulfil quite generally (see Section 4.1). The argument becomes especially dubious against the background of the unrealistically high appreciation that people have of the effects of screening programmes (see Section 5.3), since what they actually want is then something much better than what they actually get. The elementary problem, then, is that (apparent) public acceptance can be mustered by other means than good arguments. Hence, this criterion is only necessary. If we care about informed consent, the criterion should read: "The test should be acceptable to the population, on the basis of a realistic understanding of the benefits and drawbacks of the test."

---

[34] Baily and Murray (2008).

However, seen as a necessary condition, this criterion may also be understood as acknowledging the fact that screening programmes always have a public health oriented side, which implies a need to mind about the support and trust of the population. In this chapter, as well as earlier (e.g., Section 3.1.2.2 and Chapter 4), we have connected this aspect to the way in which screening programmes need to be placed in a broader societal context (for instance, by using the institutional approach suggested above). A programme that may look promising from a narrowly conceived medical and health perspective, but that would in fact serve to undercut popular trust in and support of the general systems of health care and public health, would on the whole not be a good idea, no matter what ethical basis we apply to the analysis.

7. *The natural history of the condition, including its development from latent to declared disease, should be adequately understood.*

Perhaps more than any of the other criteria, this one has been set aside in the actual practice of screening. In the wake of the human genome project, much research is now focusing on how chromosomal and genetic aberrations lead to symptoms on the phenotypic level – research areas like "functional genomics" and "proteomics". Until such research has progressed further, the *therapeutic gap*, i.e. the gap between our ability to test for the presence of genetic traits and our abilities to attend to the phenotypic expressions of these traits (i.e., to treat the diseases arising as a – partial – result of genetic mechanisms) remains. In other words, from a scientific point of view, for screening programmes targeting chromosomal and genetic disorders, the natural history is not yet adequately understood – otherwise there would be no need or room for genomic research. This, in turn, partially explains the controversial nature of prenatal screening and the new expanded neonatal screening programmes: there are no biomedical procedures qualifying as treatments for most of the conditions that can be discovered by prenatal screening[35] and many of the new conditions that are possible to detect in radically expanded neonatal screening programmes are equally untreatable or linked to treatment opportunities with rather uncertain risk-benefit profiles. This means that there is an important relation between this criterion and criterion 2.

---

[35] Regarding the idea of abortion or counselling as the treatment on offer in the case of prenatal screening, see Section 3.1.1.

There is also a relation to criterion 5. This since better understanding of the underlying biological mechanisms of disease most often facilitates better testing methods. For example, a more specific understanding of exactly which cases of elevated cancer risk or early signs of actual cancer that in fact develop into serious cancer and which do not (i.e. the natural history), could probably lead to a reduction of false negatives and positives, as well as overdiagnosis, in screening for prostate cancer.[36] Something similar can be said with regard to mammography and several other actual or suggested targets of screening.

Hence, if this criterion is thought of as necessary for screening to be warranted and if one applies a strictly *scientific* reading of "adequately understood", much screening today would not pass. However, we suppose that the reason Wilson and Jungner included the need of understanding the natural history of a disease in their criteria is that such understanding is often important in order to have acceptable treatments and suitable tests. So, perhaps it is reasonable to see this criterion as applying a somewhat more relaxed or flexible idea of what makes for an *adequate* understanding. Rather than the sort of understanding required in a strict scientific explanation, the adequacy of the understanding must be related to the degree to which the criteria 2 and 5 can be met. Even so, the failure of many modern or suggested screening programmes to live up to this criterion in a strict scientific sense helps to explain why they continue to be ethically problematic and provoke controversy.

8. *There should be an agreed policy on whom to treat as patients.*

Wilson and Jungner presumably related this criterion to determining which results should be counted as positive and which as negative, thus counting the former group as "patients". In addition, we have seen that some screening programmes may also view those testing negatively as patients – although from a purely preventive point of view. So, perhaps a better formulation of this criterion would be in terms of an agreed policy of which test results are to be counted as negative and which as positive and what follow-up procedures that should be available for these groups within the programme. In any case, this criterion is necessary for the ability of a programme to apply procedures of informed consent and adequate counselling. If it is unclear what results should be counted as positive and what as negative, there is no way of clearly explaining the meaning of the offer of entering a screening

---

[36] Justman (2010).

programme. Besides likely leading to confusion and frustration, both among health care staff and the targeted population, this implies that it becomes difficult, if not impossible, to assess the prospective benefits and burdens of the programme, even in theory.

"Agreement" can be more or less universal, however. For instance, different screening studies use different cut-off points in PSA screening (see Section 5.2). This need not be a great problem if one is consistent within studies and populations. However, health care staff must be prepared to justify why they have chosen differently than other groups to the well-informed patient.

Just as the other criteria, this one can hardly be sufficient. Moreover, even if it is a necessary condition, agreement *as such* cannot be a very strong argument for a screening programme. For example, we could imagine a cartel of DTC genetic testing companies reaching an agreement simply by calculating what cut-off points presented to possible customers as representing "high risk", would maximise joint profit over a period of time and similar examples may be imagined also with regard to public hospitals, clinics or labs, but then in terms of maximising added value in terms of resources available for other things than the actual programme (such as research). Presumably, as indicated by available evidence regarding the frequent use of misleading relative risk figures, such cut-off points would be rather low (since that is expected to boost sales or the basic resource base). In effect, for an agreement to be much of an argument in favour of screening, it needs to be supported by ethically acceptable and otherwise *good reason*. For instance, countries should have screening suggestions made by national proponents systematically reviewed by independent international expertise before considering taking any such baits.

9. *The cost of case-finding (including diagnosis and subsequent treatment of patients) should be economically balanced in relation to possible expenditure on medical care as a whole.*

Although this is a very plausible criterion, both from the point of view of efficiency and justice, it is very difficult to evaluate and fulfil for screening programmes, as we have seen. We have argued that – due to arguments in terms of both autonomy and health in favour of ambitious counselling organisations within screening programmes – this criterion, in fact, poses one of the main challenges for the justification of screening. Moreover, we have argued that the burden of proof and demonstration as to the extent to which this criterion is in fact met should rest on those who claim that it is fulfilled (Section 6.5.3).

10. *Case-finding should be a continuing process and not a "once and for all" project.*

This criterion touches on elementary ethical responsibilities of health care and health care staff, as well as of public health institutions, not to abandon people in need. This responsibility is often held out as particularly strong when it can be said to be actively assumed; when health care has already started a process of responding to the perceived health needs of people. However, the purpose of a screening programme seems to be relevant for determining exactly *how* continuous a screening programme needs to be. For instance, a screening programme intended to contain a communicable disease from causing a pandemic would need to continue only as long as there is a threat. Contrast this example with neonatal PKU-screening, which may be possible to argue should continue until we know that the mutation no longer exist in the population, i.e. within a foreseeable future. This means that ethical analysis on this point needs to accommodate itself to what is the appropriate goal of different programmes. Accordingly, the institutional approach – sensitive as it is to variations in this respect – harbours the analytical means for determining the issue of the proper degree of continuation of screening programmes.

6.7 Closing

This book is about the ethics of screening. Screening is a *way of organising* investigations, testing and follow-up procedures using the tools of health care and medicine. More specifically, with screening, we mean the use of medical testing methods at the initiative of health care or society for investigating the health status of individuals with the aim of selecting some of these for possible further treatment from a large population of people that is not united by previously recognised risk or symptoms of disease. Although this characterisation is vague, it should be clear from our discussions what our focus have been, namely medical investigations at the initiative of society that is of such a large scale as to cover, e.g., an entire age segment of the population of a country. Paradigmatic examples of this kind of screening is prenatal combined serum screening for Down syndrome, neonatal screening of the type run in most developed countries, routine developmental and general health checkup programs for children and adolescents and adult screening for prostate- or breast cancer.

Although this has been our main focus, our discussions are also relevant for screening on a smaller scale and even when there is some initial awareness of risk of disease among at least some of those screened. The basic values are the same and, often, so are the problems. The characteristic of screening that gives rise to many of these problem, e.g. regarding autonomy and arising anxiety, is that the initiative to testing and what follows comes not from the individual but from an external and quite powerful party. Hence, the discussions about these problems are relevant also for, e.g., cascade genetic testing. And, as we have seen, there are relevant similarities between screening and DTC genetic testing that allows for an illumination of the problems of the latter. Our ambition has thus been to map the ethical terrain of screening so as to isolate and illuminate the factors relevant for the assessment of any screening programme. Here follows a summary of some of our main points.

In order to provide a rationale or justification for any screening programme one has to have some idea about the values that the programme should promote. This should be the point of departure for any screening and the yardstick with which any screening programme must be evaluated. There are three kinds of basic values or goals invoked as the rationale for introducing screening programmes: improvement of health, improvement of psychological well-being or promotion of autonomy. We have analysed several conflicts within and between these goals, e.g. the potential tension between the traditional goal within health care of promoting individual health and the goal of public health to promote the health of the population, as well as the potential conflict between respecting and promoting autonomy. We have also argued that screening programmes must primarily be justified with reference to their potency for resulting in some kind of improvement of physiological health. As repeatedly argued, the very fact that the initiative does not come from the individual herself always presents some problem from the point of view of autonomy and psychological well-being. This does not mean that the other values are irrelevant, quite the contrary. Rather, the organising of something as complicated as a screening programme makes it less likely that these other values are realised to such an extent as to warrant the implementation of screening. However, to introduce the very same tests in ordinary health care for reasons of psychological well-being and autonomy can many time be justified. This basic point cannot be emphasised enough: to argue in favour of introducing a test in health care is one thing, to argue that it should be used within a screening programme is another.

So, a screening programme has to be evaluated in terms of the goals it tries to achieve and the overall balance of its costs and benefits compared to alternative measures. Several more specific factors have to be considered in order to operationalise goals, benefits, and costs. We have been identifying and

6.7 Closing

investigating these factors. They include properties of the disease in question, the tests and analytical methods applied, and the treatments applied. Since different diseases have their onset in different ages, there is also the related question of when in the patients' lives the test should be made – during the prenatal, neo-natal, childhood or adult stage.

First, the disease screened for must be an important health problem, which means that the prevalence of disease in a population is high and that the disease is serious enough for the individual. If the disease is very serious, prevalence need not be as high and reversely. The actual practice of screening oscillates between giving prevalence the decisive weight and holding out severity in the single case as their primary rationale. However, since different screening programmes can have different goals depending on who is the target of the programme this means that the assessment of the ethics of such a programme needs to proceed from different outsets depending on the target population. For instance, reproductive autonomy is foremost a goal, or, rather, an alleged goal, for prenatal screening, while other screening programmes with other targets aim for other goals. Thus, the goals can affect what conditions should be considered serious enough for screening.

However, even if one grants that reproductive autonomy is an important goal for prenatal diagnosis, it is a highly questionable argument in favour of organising this practice in the form of screening programmes, in view of the obvious risk of such an organisation of the practice to express discriminatory messages about disabled people, to undercut patient autonomy, and noting that the chief aim of these methods (to reduce the amount of invasive tests) can be attained without organising the practice as a screening.

All developed countries have more or less ambitious neonatal screening programmes. The diseases screened for in these programmes have until recently shared some common characteristics: they can be detected at this early age, the condition most likely had not been detected if not screened for and early measures are available as well as necessary for preventing or ameliorating the diseases in question. Moreover, the tests for the diseases are very reliable and safe, and the diseases themselves are very serious. Therefore, if there ever was a sunny story of screening, neonatal screening is often made out to be the one. These are, roughly, the criteria of Wilson and Jungner for screening, and we have argued, throughout the book, that these criteria still are reasonable ones for *any* screening. However, depending on how the criteria and goals are interpreted in detail, and differences with regard to socio-economical and political structure, the policy and practical recommendation for neonatal screening can vary between countries. On this basis, we have wielded some criticism against the more radical expansion of neonatal screening that has taken place in recent decades, particularly in the USA. Especially in the light of this development, we have also questioned the idea of avoiding to apply standards of counselling and informed consent to parents in this area, albeit recognising that parental *consent* may be argued as questionable in the traditional, more restrictive neonatal programmes.

Screening programmes targeting children and adolescents often come in the form of broad developmental and health checkups and such programmes are of a sort that seems quite easy to defend, since they provide clear benefits to all participants on both the individual level and that of public health. At the same time screening children and adolescents actualise some complicated ethical issues that attain particular importance in light of suggestions to screen children for conditions where the risk-benefit ratio is less obviously favourable (such as ATD or ADHD). Some of these issues afflict neonatal screening as well, although likely to a lesser extent in more restricted neonatal programmes, e.g. the issue of how to handle parents that are unwilling to have their children entering the programme, and the issue of how to handle stigmatisation as a consequence of positive test results. Some issues instead make the problems of child- and adolescent screening more similar to the problems of adult screening, e.g. the question of how to respond to the fact that children may be more or less autonomous. In relation to both childhood and adult screening, we emphasised the need for ethical assessments of screening programmes to consider a wider socio-political context. This is so, partly due to the fact that the information provided by the tests used in screening programmes may be of interest not only for health care and the individual, but also for various third parties, e.g. relatives, insurance companies, and employers.

As regards the properties of the methods for testing and analysis applied, we addressed three aspects with regard to this: safety, validity and predictive value. One of the most important conclusions from that analysis was that in order to determine what to opt for, higher sensitivity or higher specificity, a closer examination is needed of various other variables, e.g. the disease targeted, the consequences for the individuals tested as well as for society and health care in general, and the access of treatments. Another important conclusion was that, as a rule, the less predictive value, the stronger the reason to abstain to use the test in a screening programme.

The differences between ordinary health care testing and testing done within screening programmes also have consequences as regards treatment, e.g. that the general responsibility of health care to be able to offer an acceptable treatment at the outset is stronger in the screening case than in the ordinary health care situation. Moreover, we argued that a choice has to be made between having population health or reproductive autonomy as

6.7 Closing

the guiding value of prenatal screening, that it is reasonable to chose the latter, but that this speaks against organising prenatal testing as screening. Similarly, the prospect of justifying screening programmes in virtue of a follow-up procedure for those who have tested positively consisting of mere counselling does not look very bright.

Screening programmes involve the processes of contacting people for recruitment to the programme, informing them about the procedures prior to testing, as well as about the result of the test afterwards, counselling about possible follow up-procedures, and help with coping with the reactions to the test result. Even if defensible in terms of the disease targeted and the testing method utilised, a programme may still be open to serious criticism if organised in an inferior way as regards these processes.

There is a growing consensus that also screening has to apply procedures of informed consent, i.e. disclosing information about the testing and its possible consequences in a way relevant and understandable to the participant and without applying autonomy-restricting pressure to participate. This implies that it is problematic to screen for a large number of disorders simultaneously, due to the risk of information overload. This also implies that screening programmes needs to consider beforehand questions regarding secondary or unexpected information, as well as the potential conflict between promoting public health and respecting individual autonomy.

In order for individuals to benefit from the medical information resulting from a screening programme, what information is disclosed and how it is disclosed is of great importance. Counselling is a practice founded on the norm of non-directiveness that aims at designing the situation of disclosure so that it is conducive to the health, well-being and autonomy of the individual. It is not an easy affair to achieve this goal; it takes time, effort, training, and, thus, substantial resources. This presents a special problem for screening programmes, partly because they involve such large quantities of people, partly because these people are as a rule not very prepared for the prospect of being tested. The problem of benefiting and not harming the patient to a satisfying degree to an acceptable cost that this creates for screening programmes at least places the onus probandi on the one proposing such a programme: benefits must clearly outweigh harms in order for the implementation of a screening programme to be justified, and securing that so is the case must not make the programme too costly. This gives rise to difficult questions regarding the cost-effectiveness of screening programs, as well as questions of how to balance efficient uptake of participants with respect for their autonomy. Programmes involving testing for many conditions where the risk, prognosis and/or prospects for beneficial treatment is uncertain (such as the more radically expanded neonatal programmes or prenatal screening) are afflicted by this problem to a particularly large extent.

We have also presented four cases of screening and analysed them based on previous discussions. These cases were non-invasive prenatal diagnosis, neonatal screening for fragile X, mammography screening and PSA screening. These cases are of current interest, controversial, and telling regarding the general debate on which screening is defensible. The discussions showed that the factors from the previous discussions were relevant in order to take a stand on the justifiability of the screening programmes. Two other important conclusions followed from the case discussions. One was that screening exerts an institutional pull, which makes it especially important to carefully ponder them before introduction. The other was the need to sort out and make explicit ethical considerations and value judgements when discussing screening from a scientific point of view.

Although the basic values of screening are not in principle different from the basic values of health care or public health pursuits in general, there are a lot of conflicts between and within those values. How to handle these conflicts more in detail have been the subjects of this final chapter. For instance, the tension between applying a standard health care ethical or a public health ethical perspective on screening remains. A related example is the question of how to trade off severity and prevalence of the targeted disease when these factors pull in opposite directions. The situation is complicated by the fact that all screening programmes give rise to the need to consider a much wider social context than that of medical and technical facts, the individual testing situation and the immediate practicalities of running a programme. So the question arose, in the light of conflict of goals and values within screening programs, on what basis should the ethical plausibility of the goals of screening programmes be assessed?

In order to answer that question, we presented an outline of the institutional approach, which allows for different institutions to have different goals and evaluate their practices in the light of these goals. We then illustrated how this approach may be applied in three cases: prenatal care vs. communicable disease, DTC genetic testing, and the wider question of how to prioritise between screening and other ways of organising health care. Although the institutional approach allows for different institutions to have different goals, it became clear that this by no means produces a relativistic position, since the basic values of screening still holds. In fact, we argued that the classic Wilson and Jungner criteria still, to a large extent, are plausible criteria for any screening programme, if only properly interpreted.

6.7 Closing

Sometimes screening is a good idea, sometimes it is not. The question is; when is it? In this book, we have tried to address that question. This has given rise to new questions that merit further investigation. We do not pretend to have answered all questions regarding the ethics of screening. However, we have tried to sketch an outline of a theoretical framework that might be one way of approaching a comprehensive theory on the ethics of screening.

References

Ablon, J. 2002. The nature of stigma and medical conditions. *Epilepsy & Behaviour* 3(6), Supplement 2:2–9.
ACMG (American College of Medical Genetics). 2005. *Newborn screening: Toward a uniform screening panel and system*. Rockville, MD: Maternal and Health Care Bureau. Also available at URL: http://mchb.hrsa.gov/screening.
Adelswärd, V., and L. Sachs. 2002. *Framtida skuggor: Samtal om risk, prevention och den genetiska familjen* (*Future shadows: Conversations on risk, prevention and the genetic family*). Arkiv förlag: Lund.
Andermann, A., I. Blanquaert, S. Beauchamp, and V. Déry. 2008. Revisiting Wilson and Jungner in the genomic age: A review of screening criteria over the past 40 years. *Bullentin of the World Health Organization* 86(4):317–319.
Annes, J.P., M.A. Giovanni, and M.F. Murray. 2010. Risks of presymptomatic direct-to-consumer genetic testing. *New England Journal of Medicine* 363(12):1100–1101.
Asch, A. 2002. Disability, equality, and prenatal testing: Contradictory or compatible? *Florida State University Law Review* 30:315–342.
Ashcroft, R.E., A. Dawson, H. Draper, and J.R. McMillan (eds.) 2007. *Principles of health care ethics*, 2nd edition. Chichester: Wiley.
Bailey, D.B., Jr., D. Skinner, and S.F. Warren. 2005a. Newborn screening for developmental disabilities: Reframing presumptive benefit. *American Journal of Public Health* 11:1889–1893.
Bailey, D.B., Jr., D. Skinner, and K.L. Sparkman. 2005b. Discovering fragile X syndrome: Family experiences and perceptions. *Pediatrics* 2:407–416.
Bailey, D.B., Jr., D. Skinner, A.M. Davis, I. Whitmarsh, and C. Powell. 2008. Ethical, legal, and social concerns about expanded newborn screening: Fragile X syndrome as a prototype for emerging issues. *Pediatrics* 3:693–704.
Baily, M.A., and T.H. Murray. 2008. Ethics, evidence, and cost in newborn screening. *Hastings Center Report* 3:23–31.
Barry, B. 1991. *Theories of justice*. Berkeley, CA: University of California Press.
Battin, M.P., L.P. Francis, J.A. Jacobson, and C.B. Smith. 2008. *The patient as victim and vector: Ethics and infectious disease*. Oxford: Oxford University Press.
Beauchamp, T.L., and J.F. Childress. 2001. *Principles of biomedical ethics*, 5th edition. New York, NY and Oxford: Oxford University Press.
Benn, P.A., and A.R. Chapman. 2010. Ethical challenges in providing noninvasive prenatal diagnosis. *Current Opinion in Obstetrics and Gynecology* 22:128–134.

Boorse, C. 1997. A rebuttal on health. In *What is disease?*, eds. J. Humber and R. Almeder, 3–134. Totowa, NJ: Humana Press.

Brandon, K. 2009. 12.6 million is a BIG number. *The white house blog*. Online access: http://www.whitehouse.gov/blog/126-Million-is-a-BIG-Number/. Accessed 11 Aug 2009.

Broberg, G., and N. Roll-Hansen. 2005. *Eugenics and the welfare state: Sterilization policy in Denmark, Sweden, Norway, and Finland*. East Lansing, MI: Michigan State University Press.

Buchanan, A., D.W. Brock, N. Daniels, and D. Wikler. 2000. *From chance to choice – Genetics and justice*. Cambridge: Cambridge University Press.

Bui, T.H., and M. Nordenskjöld. 2002. Prenatal diagnosis: Molecular genetics and cytogenetics. *Best Practice Research in Clinical Obstetrics and Gynaecology* 16: 629–643.

Chadwick, R., D. Shickle, H. ten Have, and U. Wiesing. (eds.) 1999. *The ethics of genetic screening*. Dordrecht: Kluwer Academics Publishers.

Chalmers, D. 2005. *Genetic testing and the criminal law*. London: Routledge and Chapman & Hall.

Childress, J.F., R.R. Faden, R.D. Gaare, L.O. Gostin, J. Kahn, R.J. Bonnie, et al. 2002. Public health ethics: Mapping the Terrain. *Journal of Law, Medicine & Ethics* 30(2):170–178.

Chiu, R.W.K., R. Akolekar, Y.W.L. Zheng, T.Y. Leung, H. Sun, K.C.A. Chan, et al. 2011. Non-invasive prenatal assessment of trisomy 21 by multiplexed maternal plasma DNA sequencing: Large scale validity study. *BMJ* 342:c7401.

CIOMS. 2002. *International ethical guidelines for biomedical research involving human subjects*. Geneva: World Health Organisation.

Clarke, A. (ed.) 1994. *Genetic counselling: Practice and principles*. London: Routledge.

Clarke, A. 1998. Genetic counselling. *Encyclopaedia of applied ethics, Volume 2*. San Diego, CA: Academic Press.

Coggon, J. 2010. Does public health have a personality (and if so, does it matter if you don't like it)? *Cambridge Quarterly of Healthcare Ethics* 19:235–248.

Connor, M., and M. Ferguson-Smith. 1997. *Essential medical genetics*. Oxford: Blackwell Science Ltd.

Coutts, M.C. 1990. Maternal-fetal conflict: Legal and ethical issues. *Scope notes: Annoted bibliographies from the bioethics research library*, No. 14. Available online: http://bioethics.georgetown.edu/publications/scopenotes/sn14.pdf. Accessed 28 Oct 2010.

Daniels, N. 2007. *Just health: Meeting health needs fairly*. Cambridge: Cambridge University Press.

Danish Council of Ethics. 1999. *Screening – A report*. Copenhagen: The Danish Council of Ethics.

Davis, T.C., S.G. Humiston, C.L. Arnold, J.A. Bocchini Jr., P.F. Bass 3rd, E.M. Kennen, et al. 2006. Recommendations for effective newborn screening communication: Results of focus groups with parents, providers, and experts. *Pediatrics* 117:326–340.

Dawson, A. 2005. The determination of "best interest" in relation to childhood vaccinations. *Bioethics* 19(2):187–205.

Dawson, A. 2007. Vaccination ethics. In *Principles of health care ethics*, eds. R.E. Achcroft, A. Dawson, H. Draper, and J.R. McMillan. Chichester: Wiley.

References

Dawson, A. (ed.) 2011. *Public health ethics: Key concepts and issues in policy and practice.* Cambridge: Cambridge University Press.

Dawson, A., and M. Verweij. 2007a. The meaning of 'public' in 'public health'. In *Ethics, prevention, and public health,* eds. A. Dawson and M. Verweij. Oxford: Oxford University Press.

Dawson, A., and M. Verweij. (eds.) 2007b. *Ethics, prevention, and public health.* Oxford: Oxford University Press.

Deans, Z., and A.J. Newson. 2010. Should non-invasiveness change informed consent procedures for prenatal diagnosis? *Health Care Analysis,* 2010 Mar 9. [Epub ahead of print, doi: 10.1007/s10728-010-0146-8]

de Jong, A., W.J. Dondorp, C.E. de Die-Smulders, S.G. Frints, and G.M. de Wert. 2010. Non-invasive prenatal testing: Ethical issues explored. *European Journal of Human Genetics* 18:272–277.

Denham, J.W., R. Bender, and W.E. Paradice. 2010. It's time to depolarise the unhelpful PSA-testing debate and put into practice lessons from the two major international screening trials. *Medical Journal of Australia* 192(7), 393–396.

Dew-Hughes, D. (ed.) 2003. *Educating children with Fragile X syndrome. A multi-professional view.* London: Routledge. DnaDirect. Webportal to commercial genetic services. Online access: http://www.dnadirect.com/web/consumers. Last accessed 29 Oct 2010.

Domenighetti, G., B. D'Avanzo, M. Egger, F. Berrino, T. Perneger, P. Mosconi, et al. 2003. Women's perception of the benefits of mammography screening: Population-based survey in four countries. *International Journal of Epidemiology* 32:816–821.

Duffy, S.W., T. László, and R.A. Smith. 2002. The mammography screening trials: Commentary on the recent work by Olson and Götzcshe. *Journal of Surgical Oncology* 81:159–166 (discussion 162–166).

Eddy, D.M. 1990. Anatomy of a decision. *Journal of the American Medical Association* 263:441.

ESHG (European Society of Human Genetics). 2003. Population genetic screening programmes: Technical, social and ethical issues. Recommendations of the European Society of Human Genetics. *European Journal of Human Genetics* 11:5–7.

Esserman, L., Y. Shieh, and I. Thompson. 2009. Rethinking screening for breast cancer and prostate cancer. *JAMA* 302:1685–1692.

Evans, J.P., D.C. Dale, and C. Fomous. 2010. Preparing for a consumer-driven genomic age. *New England Journal of Medicine* 363(12):1099–1103.

Fang, F., N.L. Keating, L.A. Mucci, H.O. Adami, M.J. Stampfer, U. Valdimarsdóttir, and K. Fall. 2010. Immediate risk of suicide and cardiovascular death after a prostate cancer diagnosis: Cohort study in the United States. *Journal of the National Cancer Institute* 102(5):307–314.

Favre, R., N. Duchange, C. Vayssière, M. Kohler, N. Bouffard, M.C. Hunsinger, et al. 2007. How important is consent in maternal serum screening for Down syndrome in France? Information and consent evaluation in maternal serum screening for Down syndrome: A French study. *Prenatal Diagnosis* 3:197–205.

Fraser, F.C. 1974. Genetic counseling. *American Journal of Human Genetics* 26:636–659.

Freedman, D.A., D.B. Petitti, and J.M. Robins. 2004. On the efficacy of screening for breast cancer. *International Journal of Epidemiology* 33:43–55.

Friedman, M. 1970. The social responsibility of business is to increase its profits. *The New York Times Magazine*, September 13, 1970.

Friedman Ross, L. 2011. Mandatory versus voluntary consent for newborn screening? *Kennedy Institute of Ethics Journal* 20(4):299–328.

Genetic Integrity Act. Swedish code of statutes no 2006:351. Available online: http://www.smer.se/bazment/266.aspx

Gianroli, L., M.C. Magli, A. Feraretti, and S. Munné. 1999. Preimplantation diagnosis for aneuploidies in patients undergoing in vitro fertilization with a poor prognosis: Identification of the categories for which it should be proposed. *Fertility and Sterility* 72:837–844.

Glover, J. 1977. *Causing death and saving lives*. London: Penguin Books.

Glover, J. 2006. *Choosing children: Genes, disability, and design*. Oxford: Clarendon Press.

Gregg, A.R., and J.L. Simpson. 2002. Genetic screening for cystic fibrosis. *Obstetrics and Gynecology Clinics of North America* 29:329–340.

Gøtzsche, P.C., and K.J. Jörgensen. 2009a. Ärlig information om mammografiscreening, tack! (Honest information about mammography screening, please!). *Läkartidningen* 44:2860–2861.

Gøtzsche, P.C., and K.J. Jörgensen. 2009b. Överdiagnostik vid mammografiscreening är ett allvarligt problem (Overdiagnosis in mammography screening is a serious poblem). *Läkartidningen* 47:3180.

Gøtzsche, P.C., and M. Nielsen. 2009. *Screening for breast cancer with mammography (review)*. Copenhagen: The Cochrane Collaboration, Wiley.

Gustavson, K.-H. 1989. The prevention and management of autosomal recessive conditions. Main example: Alpha-1 antitrypsin deficiency. *Clinical Genetics* 36:327–332.

Hall, A., A. Bostanci, and C.F. Wright. 2010. Non-invasive prenatal diagnosis using cell-free fetal DNA technology: Applications and implications. *Public Health Genomics* 13:246–255.

Hampton, M.L., J. Anderson, B.S. Lavizzo, and A.B. Bergman. 1974. Sickle-cell "nondisease". *American Journal of Diseases of Children* 128:58–61.

Häyry, M., and T. Takala. 2000. Genetic ignorance, moral obligations and social duties. *Journal of Medicine and Philosophy* 1:107–113.

Hempel, C.G. 1967. *Philosophy of natural science*. Upper Saddle River, NJ and Harlow: Prentice Hall.

Hernández, E.R. 2009. What next for preimplantation genetic screening? Beyond Aneuploidy. *Human Reproduction* 24(7):1538–1541.

Hildt, E. 1999. Some reflections on the use of the term 'prevention' in reproductive medicine. In *Genetics in human reproduction*, eds. E. Hildt and S. Graumann, 243–250. Aldershot: Ashgate.

Hoedemaekers, R. 1999. Genetic screening and testing. A moral map. In *The ethics of genetic screening*, eds. R. Chadwick et al. Dordrecht: Kluwer Academic Publishers.

Hoffman, R.M. 2010. Randomized trial results did not resolve controversies surrounding prostate cancer screening. *Current Opinion in Urology* 20:189–193.

Hoffman, R.M., M.P. Couper, B.J. Zikmund-Fisher, C.A. Levin, M. McNaughton-Collins, D.L. Helitzer, et al. 2010. Prostate cancer screening decisions: Results from the National Survey of Medical Decisions (DECISIONS study). *Archives of Internal Medicine* 169:1611–1618.

References

Holm, S. 1999. There is nothing special about genetic information. In *Genetic information: Acquisition, access and control*, eds. A.R. Thompson and R. Chadwick, 97–103. New York, NY: Kluwer Academics/Plenum Publishers.

Hugosson, J., S. Carlsson, G. Aus, S. Bergdahl, A. Khatami, P. Lodding, et al. 2010. Mortality results from the Göteborg randomized population-based prostate-cancer screening trial. *Lancet Oncology* 11:725–732.

Ilic, D., D. O'Connor, S. Green, and T. Wilt. 2008. *Screening for prostate cancer (Review)*. Copenhagen: The Cochrane Collaboration, Wiley.

Infertility Resources Website. Internet portal to US commercial providers of assisted reproductive technologies and PGD. Online access: http://www.ihr.com/infertility/provider/preimplantation-genetic-diagnosis-pgd.html. Last accessed 29 Oct 2010.

Ioannou, P. 1999. Thalassemia prevention in Cyprus. Past, present and future. In *The ethics of genetic screening*, eds. R. Chadwick, et al. Dordrecht: Kluwer Academics Publishers.

Jenssen Hagerman, R., and P.J. Hagerman. (eds.) 2002. *Fragile X syndrome: Diagnosis, treatment, and research*, 3rd edition. Baltimore, MD: Johns Hopkins University Press.

Juengst, E.T. 2003. Enhancement uses of medical technology. In *Encyclopedia of bioethics*, 3rd edition, ed. S.G. Post. New York, NY: Macmillan Reference USA.

Justman, S. 2010. Uninformed consent: Mass screening for prostate cancer. *Bioethics*, Article first published online: 28 Jun 2010, doi: 10.1111/j.1467-8519.2010.01826.x.

Juth, N. 2005. *Genetic information – values and rights: The morality of presymptomatic genetic testing*. Göteborg: Acta Universitatis Gothoburgensis.

Juth, N., and C. Munthe. 2007. Screening: Ethical aspects. In *Principles of health care ethics*, 2nd edition, eds. R.E. Ashcroft, A. Dawson, H. Draper, and J.R. McMillan, 607–616. Chichester: Wiley.

Kessler, S. 1997. Psychological aspects of genetic counseling. XI. Nondirectiveness revisited. *American Journal of Medical Genetics* 72:164–171.

Kinzler, W.L., K. Morrell, and A.M. Vintzileos. 2002. Variables that underlie cost efficacy of prenatal screening. *Obstetrics and Gynecology Clinics of North America* 29:277–286.

Kligman, G. 1998. *The politics of duplicity: Controlling reproduction in Ceausescu's Romania*. Berkeley and Los Angeles, CA: University of California Press.

Kottow, M.H. 2002. Who is my brother's keeper? *Journal of Medical Ethics* 28:24–27.

Lamont, J., and C. Favor. 2007. Distributive justice. In *Stanford encyclopedia of philosophy*, ed. E.N. Zalta. Stanford, CA: Stanford University. Available online: http://plato.stanford.edu/entries/justice-distributive/. Last accessed 1 Nov 2010.

Levy, J. 1980. Vulnerable children: Parents' perspectives and the use of medical care. *Pediatrics* 65:956–963.

Linnane, E., A. Paul, and R. Parry. 1999. Screening of newborn infants for cholestatic hepatobiliary disease. Does test fulfil screening criteria? *BMJ* 319:1435.

Locke, J. 1689. *Two treaties of government*, ed. P. Laslett. New York: Cambridge University Press, 1988.

Loeber, G., D. Webster, and A. Aznarez. 1999. Quality evaluation of newborn screening programs. *Acta Paediatrica* 432:3–6.

Lun, F.M., N.B. Tsui, K.C. Chan, T.Y. Leung, T.K. Lau, P. Charoenkwan, et al. 2008. Noninvasive prenatal diagnosis of monogenic diseases by digital size selection and relative mutation dosage on DNA in maternal plasma. *Proceedings of the National Academy of Sciences of the United States of America* 50:1990–1995.

Malmqvist, E., G. Helgesson, J. Lehtinen, K. Natunen, and M. Lehtinen. 2010. The ethics of implementing human papillomavirus vaccination in developed countries. *Medicine, Health Care, and Philosophy* 14:19–27.

March, J.G., and J.P. Olsen. 1989. *Rediscovering institutions: The organizational basis of politics*. New York, NY: Free Press.

Marteau, T., and M. Richards. (eds.) 1996. *The troubled helix: Social and psychological implications of the new human genetics*. Cambridge: Cambridge University Press.

McNamee, M., A. Müller, I. van Hilvoorde, and S. Holm. 2009. Genetic testing and sports medicine ethics'. *Sports Medicine* 39(5):339–344.

McPherson, K. 2010. Should we screen for breast cancer? *BMJ* 341:233–234.

Meade, N., and P. Bower. 2000. Patient-centeredness: A conceptual framework and review of the empirical literature. *Social Science & Medicine* 51:1087–1110.

Mill, J.S. 1859. *On liberty*. Suffolk: Penguin Books, 1974.

Mohamed, K., R. Appleton, and P. Nicolaides. 2000. Delayed diagnosis of Duchenne muscular dystrophy. *European Journal of Paediatric Neurology* 4(5):219–223.

Moran, N.E., D. Shickle, C. Munthe, K. Dierickx, C. Petrini, F. Piribauer, et al. 2006. Are compulsory immunisation and incentives to immunise effective ways to achieve herd immunity in Europe? In *Ethics and infectious disease*, eds. M. Selgelid and M. Battin. London: Blackwell.

Munthe, C. 1996. *The moral roots of prenatal diagnosis. Ethical aspects of the early introduction and presentation of prenatal diagnosis in Sweden*. Göteborg: Centrum för forskningsetik.

Munthe, C. 1999. *Pure selection. The ethics of preimplantation genetic diagnosis and choosing children without abortion*. Göteborg: Acta Universitatis Gothoburgensis.

Munthe, C. 2005. Ethical aspects of controlling genetic doping. In *Genetic technology and sport: Ethical questions*, eds. C. Tamburrini and T. Tännsjö, 107–125. London and New York, NY: Routledge.

Munthe, C. 2007. Preimplantation genetic diagnosis: Ethical aspects. In *Encyclopedia of life sciences*. Chichester: Wiley.

Munthe, C. 2008. The goals of public health: An integrated, multidimensional model. *Public Health Ethics* 1(1):39–53.

Munthe, C., L. Sandman, and D. Cutas. 2011. Person centred care and shared decision making: Implications for ethics, public health and research. *Health Care Analysis* 19, online first. doi: 10.1007/s10728-011-0183-y.

Munthe, C., J. Wahlström, and S. Welin. 1998. Fosterdiagnostikens moraliska rötter. Goda handikappomsorger avgörande för den etiska kvaliteten (The moral roots of prenatal diagnosis. Good disability care crucial for the ethical quality). *Läkartidningen (Physician's Review)* 95:750–753.

Murray, C.J.L., J.A. Salomon, C.D. Mathers, and A.D. Lopez. 2002. *Summary measures of population health: Concepts, ethics, measurement and applications*. Geneva: WHO.

Myriad Genetics Website. Online access: http://www.myriadtests.com/index.php?page_id=80. Last accessed 29 Oct 2010.

Nadel, A.S., and M.L. Likhite. 2009. Impact of first-trimester aneuploidy screening in a high-risk population. *Fetal Diagnosis and Therapy* 26(1):29–34.

References

Natowicz, M. 2005. Newborn screening – setting evidence-based policy for protection. *New English Journal of Medicine* 9:867–870.

Neal, D.E. 2010. PSA testing for prostate cancer improves survival – but can we do better? *The Lancet Oncology* 11:702–703.

Newschaffer, C.J., K. Otani, M.K. McDonald, and L.T. Penberthy. 2000. Causes of death in elderly prostate cancer patients and in a comparison nonprostate cancer cohort. *Journal of the National Cancer Institute* 92:613–621.

Nijsingh, N. 2007. Informed consent and the expansion of newborn screening. In *Ethics, prevention and public health*, eds. A. Dawson and M. Verweij. Oxford: Oxford University Press.

Nilsson, T., C. Munthe, C. Gustavson, A. Forsman, and H. Anckarsäter. 2009. The precarious practice of forensic psychiatric risk assessment. *International Journal of Law and Psychiatry* 32:400–407.

Nordenfelt, L. 2003. On the evolutionary concept of health: Health as natural function. In *Dimensions of health and health promotion*, eds. L. Nordenfelt and P.E. Liss, 37–56. Amsterdam: Rodopi Press.

Nozick, R. 1974. *Anarchy, state and Utopia*. New York, NY: Basic Books.

The Nuffield Council on Bioethics. 1993. *Genetic screening and ethical issues*. London: The Nuffield Council on Bioethics.

The Nuffield Council on Bioethics. 2010. *Medical profiling and online medicine: The ethics of 'personalised healthcare' in a consumer age*. Nuffield Council on Bioethics, London. Available online: http://www.nuffieldbioethics.org/personalised-healthcare-0. Last accessed 31 Oct 2010.

Pandor, A., J. Eastman, C. Beverley, J. Chilcott, and S. Paisley. 2004. Clinical effectiveness and cost-effectiveness of neonatal screening for inborn errors of metabolism using tandem mass spectrometry: A systematic review. *Health Technology Assessment* 8:1–134.

Parens, E., and A. Asch. 2000. *Prenatal testing and disability rights*. Washington, DC: Georgetown University Press.

Parker, G. 1983. *Parental overprotection: A risk factor in psychosocial development*. New York, NY: Grune & Straton.

Persad, G., A. Wertheimer, and E.J. Emanuel. 2009. Principles for allocation of scarce medical interventions. *Lancet* 373:423–431.

Platt Walker, A. 1998. The practice of genetic counselling. In *A guide to genetic counselling*, ed. D.L. Baker, 1–20. Chichester: Wiley-Liss Inc.

Post, S.G., J.R. Botkin, and P. Whitehouse. 1992. Selective abortion for familial Alzheimer disease? *Obstetrics and Gynaecolcology* 79:794–798.

The President's Council on Bioethics. 2008. *The changing moral focus of newborn screening: An ethical analysis by the President's council on bioethics*. Washington, DC: The President's Council on Bioethics.

Radetzki, M., M. Radetzki, and N. Juth. 2003. *Genes and insurance: Ethical, legal and economic issues*. Cambridge: Cambridge University Press.

Ravitsky, V. 2009. Non-invasive prenatal diagnosis: An ethical imperative. *Nature Reviews Genetics* 10:733.

Rawls, J. 1972. *A theory of justice*. London: Oxford University Press.

Renner, I. (ed.) 2006. Experience of pregnancy and prenatal diagnosis. *Bundeszentrale für gesundheitliche Aufklärung* [online] http://www.bzga.de/?uid=25d093aacb9296ea646b087b68c27996&id=medien&sid=88&idx=1496. Accessed Nov 2010.

Rhodes, R. 1998. Genetic links, family ties and social bonds: Rights and responsibilities in the face of genetic knowledge. *Journal of Medicine and Philosophy* 23:10–30.

Rhodes, R. 2005. Justice in medicine and public health. *Cambridge Quarterly of Healthcare Ethics* 14(1):13–26.

Robertson, S., and J. Savulescu. 2001. Is there a case in favour of predictive genetic testing in young children? *Bioethics* 15:26–49.

Rose, G. 1992. *The strategy of preventive medicine*. Oxford: Oxford University Press.

Rosen, G. 1993. *A history of public health, expanded edition*. Baltimore, MD: Johns Hopkins University Press.

Rowland, A.S., D.M. Umbach, K.E. Catoe, S. Long, D. Rabiner, A.J. Naftel, R. Panke Faulk, and D.P. Sandler. 2001. Studying the epidemiology of attention-deficit hyperactivity disorder: Screening method and pilot results. *Canadian Journal of Psychiatry* 46:931–940.

Saltvedt, S. 2005. *Prenatal diagnosis in routine antenatal care – A randomised controlled trial*. Stockholm: Kongl Carolinska Medico Chirurgiska Institutet.

Salwén, H. 2003. *Hume's law: An essay on moral reasoning*. Stockholm: Almqvist & Wiksell International.

Sandén, M.-L., and P. Bjurulf. 1988. Pregnant women's attitudes for accepting or declining a serum-alpha-fetoprotein test. *Scandinavian Journal of Social Medicine* 16:265–271.

Sandman, L., and C. Munthe. 2009. Shared decision making and patient autonomy. *Theoretical Medicine and Bioethics* 30(4):289–310.

Sandman, L., and C. Munthe. 2010. Shared decision making, paternalism and patient choice. *Health Care Analysis* 18(1):60–84.

Savulescu, J. 2005. Compulsory genetic testing for APOE Epsilon 4 and boxing. In *Genetic technology and sport: Ethical questions*, eds. C. Tamburrini and T. Tännsjö, 136–146. London and New York, NY: Routledge.

Schmitz, D., C. Netzer, and W. Henn. 2009. An offer you can't refuse? Ethical implications of non-invasive prenatal diagnosis. *Nature Reviews Genetics* 10:515.

Schröder, F.H., J. Hugosson, M.J. Roobol, T.L. Tammela, S. Ciatto, V. Nelen, et al. 2009. Screening and prostate-cancer mortality in a randomized European study. *The New England Journal of Medicine* 13:1320–1328.

Scott, R. 2005. Prenatal testing, reproductive autonomy, and disability interests. *Cambridge Quarterly of Healthcare Ethics* 14:65–82.

Segall, S. 2010. *Health, luck, and justice*. Princeton, NJ: Princeton University Press.

Shattuck-Eidens, D., et al. 1997. BRCA1 Sequence analysis in women at high risk for susceptibility mutations. *Journal of the American Medical Association* 15: 1242–1250.

Shickle, D. 1999. The Wilson and Jungner principles of screening and genetic testing. In *The ethics of genetic screening*, eds. R. Chadwick et al. Dordrecht: Kluwer Academics Publishers.

Shickle, D., and I. Harvey. 1993. Inside-out, back-to-front: A model for clinical population genetic screening. *Journal of Medical Genetics* 30:580–582.

Shickle, D., E. Richardson, F. Day, C. Munthe, A. Jovell, H. Gylling, et al. 2007. *Public policies, law and bioethics: A framework for producing public health policy across the European Union*. Leeds: University of Leeds. Online access: http://www.leeds.ac.uk/lihs/ihsphr_ph/documents/EurophenFullReport.pdf. Accessed 25 Oct 2010.

Shiloh, S. 1996. Decision-making in the context of genetic risk. In *The troubled helix: Social and psychological implications of the new human genetics*, eds. T. Marteau and M. Richards, 82–103. Cambridge: Cambridge University Press.

Skinner, D., K. Sparkman, and D.B. Bailey Jr. 2003. Screening for fragile X syndrome: Patient attitudes and perspectives. *Genetics in Medicine* 5:378–384.

Skrabanek, P. 1985. False premises and false promises of breast cancer screening. *The Lancet* 326(8446):94–95.

Skrabanek, P. 2000. *False premises, false promises: Selected writings of Peter Skrabanek*. Glasgow: Tarragon Press.

Smith, D.S., P.A. Humphrey, and W.J. Catalona. 1997. The early detection of prostate carcinoma with prostate specific antigen: The Washington University experience. *Cancer* 80:1853–1856.

Smith Iltis, A. 2001. Organizational ethics and institutional integrity. *HEC Forum* 13(4):317–328.

Sobel, S.K., and D.B. Cowan. 2000. Impact of genetic testing for Huntington disease on the family system. *American Journal of Medical Genetics* 90:49–59.

State of Michigan. 1978. *Public health code, Act 368 of 1978*. Lansing, MI: The State of Michigan.

Stein, R. 2010a. Company plans to sell genetic testing kit at drugstores. *Washington Post*, 11 May 2010. Available online: http://www.washingtonpost.com/wp-dyn/content/article/2010/05/10/AR2010051004904.html. Last accessed 29 Oct 2010.

Stein, R. 2010b. Walgreens won't sell over-the-counter genetic test after FDA raises questions. *Washington Post*, 13 May 13 2010. Available online: http://www.washingtonpost.com/wp-dyn/content/article/2010/05/12/AR2010051205156.html. Last accessed 29 Oct 2010.

Sutton, V.R. 2002. Tay-Sachs disease. Screening and counseling families at risk for metabolic disease. *Obstetrics and Gynecology Clinics of North America* 29:287–296.

Swedish Organized Service Screening Evaluation Group. 2006. Reduction in breast cancer mortality from organized service screening with mammography. Further confirmation with extended data. *Cancer Epidemiology, Biomarkers and Prevention* 15:45–51.

ten Have, H. 2000. Genetics and culture. In *Bioethics in a European perspective*, eds. H. ten Have and B. Gordijn. Nijmegen: Programme of the European Masters in Bioethics.

Thelin, T., T. Sveger, and T.F. McNeil. 1996. Primary prevention in a high-risk group: Smoking habits in adolescents with homozygous alpha-1-antitrypsin deficiency (ATD). *Acta Paediatrica* 85:1207–1212.

Thomasgard, M. 1998. Parental perception of child vulnerability, overprotection, and psychological characteristics. *Child Psychiatry and Human Development* 28:223–240.

Tännsjö, T. 1999. *Coercive care. The ethics of choice in health and medicine*. London: Routledge.

Törnberg, S., and L. Nyström. 2009a. Skrämselpropaganda om mammografi (Propaganda with intention to scare regarding mammography). *Läkartidningen* 42:2664–2665.

Törnberg, S., and L. Nyström. 2009b. Värre med utredningsorsakad oro än förtidig död i bröstcancer? (Is it worse with examination-induced anxiety than premature death in cancer?). *Läkartidningen* 45:3018.

U.S. Preventive Services Task Force. 2008. Screening for prostate cancer: US. Preventive Services Task Force recommendation statement. *Annals of Internal Medicine* 149:185–191.

Vahab Saadi, A., P. Kushtagi, P.M. Gopinath, and K. Satyamoorthy. 2010. Quantitative fluorescence polymerase chain reaction (QF-PCR) for prenatal diagnosis of chromosomal aneuploidies. *International Journal of Human Genetics* 10(1–3):121–129.

van den Heuvel, A., L. Chitty, E. Dormandy, A. Newson, Z. Deans, S. Attwood, S. Haynes, and T.M. Marteau. 2009. Will the introduction of non-invasive prenatal diagnostic testing erode informed choices? An experimental study of health care professionals. *Patient Education and Counseling* 78:24–28.

van der Schoot, C.E., S. Hahn, and L.S. Chitty. 2008. Non-invasive prenatal diagnosis and determination of fetal Rh status. *Seminars in Fetal & Neonatal Medicine* 13, 63–68.

Verweij, M. 2000. *Preventive medicine between obligation and aspiration*. Dordrecht, Boston, MA and London: Kluwer Academic Publishers.

Wakefield, J.C. 1992. The concept of mental disorder. On the boundary between biologocial facts and social values. *American Psychologist* 4(3):373–388.

Walzer, M. 1983. *Spheres of justice. In defence of pluralism and equality*. New York, NY and London: Basic Books.

WHO. 1998. *Proposed international guidelines on ethical issues in medical genetics and the provision of genetic services*. Geneva: WHO 15–16 Dec 1997.

Wilcken, B. 2009. Cystic fibrosis: Refining the approach to newborn screening. *The Journal of Pediatrics* 155(5):605–606.

Wilcken, B., V. Wiley, J. Hammond, and K. Carpenter. 2003. Screening newborns for inborn errors of metabolism by tandem mass spectrometry. *The New England Journal of Medicine* 23:2304–2312.

Wilkinson, S. 2003. *Bodies for sale. Ethics and exploitation in the human body trade*. London: Routledge.

Wilson, J. 2009. Towards a normative framework for public health ethics and policy. *Public Health Ethics* 2(2):184–194.

Wilson, J.M.G., and G. Jungner. 1968. Principles and practice of screening for disease. *Public Health Papers*, WHO No. 34, Geneva.

World Medical Association. 1964–2008. *Declaration of Helsinki: Ethical principles for medical research involving human subjects*. Available online: http://www.wma.net/en/30publications/10policies/b3/17c.pdf. Accessed 25 Oct 2010.

Wright, C.F., and H. Burton. 2008. The use of cell-free fetal nucleic acids in maternal blood for non-invasive prenatal diagnosis. *Human Reproduction Update* 1:139–151.

Wright, C.F., and L.S. Chitty. 2009. Cell-free fetal DNA and RNA in maternal blood: Implications for safer antenatal testing. *BMJ* 339:b2451. doi: 10.1136/bmj.b2451.

Zackrisson, S., I. Andersson, L. Janzon, J. Manjer, and J.P. Garne. 2006. Rate of overdiagnosis of breast cancer 15 years after end of Malmö mammographic screening trial: Follow-up study. *BMJ* 332:689–692.

Zamerowski, S.T., M.A. Lumley, R.A. Arreola, K. Dukes, and L. Sullivan. 2001. Favorable attitudes toward testing for chromosomal abnormalities via analysis of fetal cells in maternal blood. *Genetics in Medicine* 3:301–309.

Index

A

Ablon, J, 55
Abortion, 17, 34–38, 40, 65, 73–76, 104–105, 141, 154
ACMG (American College of Medical Genetics), 43, 49–51, 60, 110, 112
Adelswärd, V, 68
ADHD (attention-deficit/hyperactivity disorder), 55, 160
Adolescent screening, 53–58, 75–77, 87, 99, 160
Adverse selection, 59
AFP (amniotic fluid α-fetoprotein), 64–65
Ahlenius, H, vi
Alzheimer's disease, 68
Amniocentesis, 35, 61, 100
Andermann, A, 3
Annes, JP, 142
Asch, A, 17, 36–39, 83
Ashcroft, RE, v
ATD (Alpha-1 Antitrypsin Deficiency), 76–77, 99, 128, 152, 160
Autonomy, 7, 10, 13, 17–18, 21–29, 33–34, 36–37, 39–40, 42, 44–46, 48–49, 54, 56–58, 60, 64, 66–67, 69, 73–78, 80, 82–89, 91–98, 103–108, 110–111, 114, 121, 127–130, 132–134, 136, 138, 140, 148, 150, 152, 156, 158–161

B

Bailey, DB, 108, 110–113
Baily, MA, 110–111, 153
Barry, B, 147
Battin, MP, 21

Beauchamp, TL, 23–25, 38, 44, 84, 86, 125
Benn, PA, 107
Biotinidase deficiency, 43, 45
Bjurulf, P, 65
Björck, E, vi, 112
Boorse, C, 150
Bower, P, 17
Brandon, K, 60
Broberg, G, 42
Breast cancer, 5–6, 58, 61, 65–66, 71, 105, 114–116, 118–120, 157
Brülde, B, vi
Buchanan, A, 34, 38, 132
Bui, TH, 35–36
Burton, H, 100–101, 103

C

Cancer, 5–6, 58, 61, 65–68, 71, 88, 90, 100, 105, 114–120, 122–126, 155, 157
Cardiac disease, 66
Cardio-vascular disease, 68
Carrier detection, 33, 62, 84
Cascade genetic testing, 9, 158
cffNA (cell-free nucleic acid), 100, 103
Chadwick, R, 8–9, 33, 43
Chalmers, D, 3
Chapman, AR, 107
Childress, JF, 2, 23–25, 38, 44, 84, 86, 125, 130
Chiu, RWK, 101
CIOMS (Council for international organizations of medical sciences), 130

175

Chromosomal (aberration, condition, disease, or disorder), 9, 16, 33, 35, 40, 62, 72, 100, 102, 107, 111–112, 154
Clarke, A, 26, 65
Cochrane collaboration, 114–115, 120, 123
Coggon, J, 133
Communicable disease, 9, 20, 66, 86, 128, 131, 134, 137–141, 157, 162
Confidentiality, 91
Congenital hypothyroidism, 43
Connor, M, 32, 61, 68, 108
Consequentialist, 10
Cost-benefit analysis, 10
Counselling, 5, 10–11, 17, 19–20, 24, 27, 33, 47–48, 52, 54, 57–58, 60, 67, 73, 75–77, 79–81, 84–85, 87–95, 97–98, 101, 103–105, 110, 112–114, 121–122, 133, 140–141, 145, 151–156, 160–161
Coutts, MC, 140
Cowan, DB, 90
Cutas, D, vi
Cut-off point, 56, 102, 122, 156
CVS (chorionic villus sampling), 35, 61, 100
Cystic fibrosis, 9, 16, 62, 71, 84, 100, 111

D

Daniels, N, 147
Danish council of ethics, 7, 96
Davis, TC, 112
Dawson, A, 1–2, 26, 46, 54, 99
Deans, Z, 108
Declaration of Helsinki, 130
de Jong., A, 102, 106
Denham, JW, 67
Dew-Hughes, D., 110
Diabetes, 18, 66, 68, 70, 72
Diagnostic odyssey, 110–111, 113
Diagnosis, 4, 6, 8, 10, 17, 21–22, 27, 33–41, 49, 61–63, 65–66, 76, 78, 85, 90, 99–108, 110–111, 115–118, 120, 122–124, 140–142, 151–156, 159, 162
Draper, H., v

DTC (direct to consumer) genetic testing, 137, 141–147, 156, 158, 162
Down syndrome, 107–108, 157
Duchenne muscular dystrophy, 43
Duffy, SW, 114, 119

E

Eddy, DM, 17
Employers, 59, 79, 129, 160
ERSPC (the European randomized study of screening for prostate cancer), 123–124
ESHG (European Society of Human Genetics), 9, 82, 85, 88
Esserman, L, 117
Evans, JP, 142

F

Fang, F, 124
Favor, C, 147
Favre, R, 105
FDA (food and drug administration), 143–144
Ferguson-Smith, M, 32, 61, 68, 108
Financial incentives, 81
FMR-1 gene, 109
Fragile X, 5, 32, 52, 62, 71, 99, 108–114, 152, 162
Fraser, FC, 87–89
Freedman, DA, 114
Friedman, M, 144–146
Friedman Ross, L, 47, 131
Funding, 70, 81, 95–97
FXTAS, 109

G

Genetic carrier screening, 8
Genetic counselling, 84, 87–93, 113, 140, 145
Genetic disease, 16, 66, 71, 92, 102, 111
Genetic integrity act, 61
Genetics, 2, 7, 35, 43, 87–93, 113, 146
Gianroli, L, 63
Glover, J, 23, 132
Gregg, AR, 9, 71
Gustavson, KH, 76
Gøtzsche, PC, 67, 114–119, 121

Index

H
Hall, A, 106–107
Hampton, ML, 20
Harvey, I, 9
Häyry, M, 84
Harm, 23, 32, 44–45, 58, 65, 86, 117–118
Health insurance, 56, 59–60, 84
Helgesson, G, vi
Hempel, CG, 140
Hermerén, G, vi
Hernández, ER, 63
Hildt, E, 35
HIV (human immunodeficiency virus), 9, 15, 20, 88
Hoedemaekers, R, 7, 10, 24, 26–27, 82–84
Hoffman, AJ, vi, 45
Hoffman, RM, 123, 125
Holm, S, 88
HPV (human papillomavirus), 99
Hugosson, J, 124

I
Ilic, D, 123
Iatrogenic abortion, 101
Infectious disease, 2, 21, 139–140
Infertility resources website, 145
Informed consent, 23, 25, 45–49, 54, 79, 81–87, 97–98, 105–106, 108, 111–112, 121, 125, 127, 131–134, 152–153, 155, 160–161
Institutional approach (to health-care ethics), 6, 135–150, 154, 157, 162
Institutional integrity, 136
Invasive testing, 105–106
Invasive prenatal testing, 62, 104–105
Insurance companies, 59, 79, 160
Ioannou, P, 9, 33
IVF (in vitro fertilization), 63, 142, 145

J
Jenssen Hagerman, R, 110
Juengst, ET, 111
Jungner, G, 3–6, 9–10, 14, 27, 31–32, 44, 60, 63, 66, 73, 115, 117, 122, 129, 149–157, 159, 162

Justice, 7, 13, 27–29, 96, 138, 147–149, 152, 156
Justman, S, 96, 117, 125, 155
Juth, N, 17, 22–25, 59–60, 70, 84–86, 88, 91, 104–105, 113–114
Jörgensen, KJ, 114, 117, 119, 121

K
Kessler, S, 91–92
Kinzler, WL, 8, 10
Kligman, G, 141
Klinefelter's, 33, 112
Kottow, MH, 132

L
Lamont, J, 147
Levy, J, 55
Life expectancy, 14, 18, 97
Life insurance, 56–57, 59–60, 66, 79, 84, 113, 160
Life-style, 16–18, 66–67, 75–76
Likhite, ML, 62
Linnane, E, 3
Locke, J, 23
Loeber, G, 43
Lun, FM, 101
Lynöe, N, vi

M
Malmqvist, E, 99
Mammography, 5, 58, 61, 66–67, 97, 99, 114–122, 124–126, 129, 132, 155, 162
Mammography screening, 5, 58, 61, 99, 114–122, 124, 129, 132, 162
March, JG, 120
Marteau, T, 9
Mastectomy, 15, 65, 116
McNamee, M, 3
McPherson, K, 117
McMillan, J, v
Medical ethics, 1–2, 17–18, 24, 38, 40, 45, 50, 56, 91, 134
Medicalise, 2
Medicalization, 46
Mental disability, 32, 108
Mill, JS, 23
Miscarriage, 61–62, 100, 102, 104–105

Mohamed, K, 43
Monogenetic, 16, 71, 101–102, 109, 111
Moran, NE, 54
Multifactorial, 68, 70–72, 111
Munthe, C, 2–3, 17, 22, 24–26, 28, 35–37, 39, 56, 63, 74, 82–84, 86, 91, 93, 95, 107, 130, 140
Murray, CJL, 14
Murray, TH, 110–111, 153
Myriad genetics website, 146

N
Nadel, AS, 62
Natowicz, M, 50
Neal, DE, 122, 124
Newschaffer, CJ, 119
Nijsingh, N, 43, 46–48, 54, 86
NIPD (non-invasive prenatal diagnosis), 100–108, 162
Neonatal screening, 5, 8, 16, 42–54, 66, 71, 78–79, 85–86, 94–95, 97, 99, 108–114, 122, 131–132, 137, 140, 145, 152, 154, 157, 159–160, 162
Newson, AJ, 108
Non-directiveness, 91–93, 98, 161
Nordenfelt, L, 150
Nordenskjöld, M, 36
Normalization, 106
Nozick, R, 23
Nuffield council of bioethics, 143
Nyström, L, 114–115, 117–119, 121

O
Obstetric ultrasound, 8, 37, 40, 62, 100
Overdiagnosis, 116–118, 120, 122, 124, 153, 155
Overtreatment, 61, 66–67, 72, 116–122, 124, 153

P
Pandemics, 2, 21, 26, 128, 138–139, 151, 157
Pandor, A, 49, 51
Parens, E, 17, 36–39, 83
Parkinson's, 109
Pedigree, 89
Penetrance, 70, 71, 109

Persad, G, 147
PGD (preimplantation genetic diagnosis), 63, 142, 145
PKU (phenylketonuria), 6, 16, 32, 43, 45, 51, 66, 72, 99, 109, 111–112, 143
Platt Walker, A, 84, 88–90
Post, SG, 32
Prenatal screening, 9, 21, 26, 33–42, 44, 47, 49, 52, 57–58, 62, 65, 72–76, 78, 80–83, 86–87, 101–108, 127, 129, 132, 137–138, 141, 154, 159, 161–162
President's council on bioethics, 45
Presymptomatic genetic testing, 55, 145
Prevalence, 32–33, 35–36, 39, 44, 50, 52, 55, 65–66, 78, 107–108, 127, 138–139, 150–151, 159, 162
Privacy, 91
Prostate cancer, 5, 58, 66–67, 117, 122–126
Population, 1, 3, 5–11, 14–15, 20–22, 25–26, 29, 32–33, 41–42, 46, 53–54, 56–57, 60–61, 66, 68, 70–71, 74–78, 80, 85–87, 94–95, 97, 112, 114–115, 117, 123, 127–132, 134, 137–140, 143, 147, 151, 153–154, 156–160
PSA (prostate-specific antigene), 5, 58, 66–67, 99, 117, 122–126, 156, 162
PSA screening, 5, 99, 117, 122–126, 156, 162
Psychological well-being, 17–18, 29, 66, 85, 96, 158
Public health, 1–3, 7–8, 10, 14–15, 20–22, 25–29, 32, 35–36, 39, 41, 46–47, 49, 52–53, 56, 58, 60, 65–67, 74, 76, 86–87, 95–98, 107, 127–135, 137, 139–141, 145, 147, 152, 154, 157–158, 160–162
Public health ethics, 2, 96, 129–130, 135, 137, 145

Q
QF-PCR, 35–36, 39, 106
Quality of life, 5, 18, 38, 44, 60, 123

R
Radetzki, M, 59–60
Ravitsky, V, 101, 103–104

Rawls, J, 27, 147
Real cancer (invasive cancer), 115
Renner, I, 105
Rhodes, R, 2, 24
Richards, M, 9
Risk, absolute, 69, 86, 146
Risk, relative, 69–70, 121, 146, 156
Robertson, S, 53
Roll-Hansen, N, 42
Rose, G, 15
Rosen, G, 139
Rowland, AS, 55

S
Sachs, L, 9, 35, 68
Saltvedt, S, 62, 81, 104
Salwén, H, 119
Sanctity of life (the principle of), 36
Sandén, ML, 65
Sandman, L, 17, 22, 24, 26, 86, 93, 95
Savulescu, J, 3, 53
SCAD (short-chain acyl-CoA dehydrogenase deficiency), 51
Schizophrenia, 68
Schmitz, D, 40, 105
Schröder, FH, 123
Scott, R, 38
Screening, concept of, 5–11, 142
Screening of children, 19, 53, 56–57, 79
Segall, S, 106
Severity of disease, 32
Shattuck-Eidens, D, 71
Shickle, D, 2, 7–10, 17, 27, 60, 68, 130
Shiloh, S, 68, 92
Sickle cell anemia, 9, 20, 65
Simpson, JL, 9, 71
Sjöstrand, M, vi
Skinner, D, 109
Skrabanek, P, 114
Smith, DS, 124
Smith Iltis, A, 136
Sobel, SK, 90
Social science perspective, 6, 132–134, 148
Specification creep, 106–107
State of Michigan, 47

Stein, R, 143
Sterilisation, 42, 141
Stigmatisation, 38, 46, 52, 55–57, 66, 76–77, 79, 107, 128, 160
Sutton, VR, 9
Swedish organized service screening evaluation group, 118

T
Takala, T, 84
Tandem mass spectrometry, 49–50, 112
Tay-Sachs, 9
Ten Have, H, 91, 94
Thalassemia, 9, 33
Thelin, T, 77
Thomasgard, M, 55
Treatment, 4, 6–7, 10–11, 14–21, 23, 31, 33, 40, 43, 47, 49–50, 52–53, 55–56, 59, 61, 65–67, 70, 72–77, 80–81, 85, 88, 90, 92, 98, 110, 115–124, 127, 135, 150–155, 157, 159–161
Triple X syndrome, 112
Turner, 33, 112
Törnberg, S, 114–115, 117–119, 121

V
Vahab Saadi, A, 35
van den Heuvel, A, 106
van der Schoot, CE, 101
Verweij, M, 1–2, 15–16, 26, 46
von Döbeln, U, 50–51

W
Wahlström, J, vi
Wakefield, JC, 150
Walzer, M, 134, 144, 146
Weisstub, D, v
Well-being, 10, 13, 17–22, 24, 27–29, 45, 57, 66, 72, 85–87, 92–93, 96, 98, 104, 150, 152, 158, 161
Werdnig-Hoffman disease, 35
Wessel, M, vi
WHO (World health organization), 3, 88, 117
Wilcken, B, 43, 49–50, 71
Wilkinson, S, 83

Wilson, JMG, 3–6, 9–10, 14, 27, 31–32, 44, 60, 63, 66, 73, 115, 117, 122, 129–130, 149–157, 159, 162
Wilson and Jungner criteria, 3–5, 117, 129, 149–157, 162
World medical association, 125, 130
Wright, CF, 100–101, 103

X
X-linked disorder, 109, 112
XXY, 112

Z
Zackrisson, S, 116
Zamerowski, ST, 106

CPSIA information can be obtained at www.ICGtesting.com
Printed in the USA
LVOW080419161111
255201LV00004B/20/P